Live Audio

Live Audio

The Art of Mixing a Show

Dave Swallow

Amsterdam • Boston • Heidelberg • London • New York • Oxford • Paris • San Diego
San Francisco • Singapore • Sydney • Tokyo

Focal Press is an imprint of Elsevier

Focal Press is an imprint of Elsevier
30 Corporate Drive, Suite 400, Burlington, MA 01803, USA
The Boulevard, Langford Lane, Kidlington, Oxford, OX5 1GB, UK

Notices
Knowledge and best practice in this field are constantly changing. As new research
and experience broaden our understanding, changes in research methods, professional
practices, or medical treatment may become necessary.

Practitioners and researchers must always rely on their own experience and knowledge
in evaluating and using any information, methods, compounds, or experiments
described herein. In using such information or methods they should be mindful of
their own safety and the safety of others, including parties for whom they have a
professional responsibility.

To the fullest extent of the law, neither the Publisher nor the authors, contributors, or
editors, assume any liability for any injury and/or damage to persons or property as a
matter of products liability, negligence or otherwise, or from any use or operation of
any methods, products, instructions, or ideas contained in the material herein.

Library of Congress Cataloging-in-Publication Data
Application submitted

British Library Cataloguing-in-Publication Data
A catalogue record for this book is available from the British Library.

ISBN: 978-0-240-81604-3

For information on all Focal Press publications
visit our website at www.elsevierdirect.com

10 11 12 13   5 4 3 2 1

Printed in the United States of America

Working together to grow
libraries in developing countries

www.elsevier.com | www.bookaid.org | www.sabre.org

ELSEVIER   BOOK AID International   Sabre Foundation

For Finn

Contents

Prelude

After leaving school at the age of 16, I didn't really know what I was going to do with my life. I knew I liked music; I'd been playing bass in a band since I was about 14. I'd spent a lot of time taping live gigs off the radio, trying not to get any talking, and I'd spent a lot of money buying gear and going to shows. Meanwhile, my mate Mike had engineered my band for quite some time and had shown me the ropes on the odd occasion. I must admit that I quite enjoyed being behind the scenes.

I spent the summer after my last exams bumming around at a college studying some bizarre computer course. That was fun, and I can still program in Hexadecimal, but it's pretty much useless to me these days. That summer turned into a year, and by the following summer I'd found a sound engineering course up in London. It was only three days a week, and I managed to get an interview. My mum smartened me up, gave me £20, and I headed off to London town. After some normal questions about who I was, what I did, and why I wanted to be a sound engineer, I was presented with the question that would decide my fate: "Do you know what DI stands for?" At first I misheard the question and started to tell the interviewer what it was used for—my career as an engineer nearly ended there. Luckily for me, she repeated the question, and my answer was good enough . . . I was in.

After being accepted for the course, I returned to my hometown of Southend-on-Sea and went to the local PA firm, Maple Studios, which also owned the local rehearsal studios, local venue, and eventually one of the local recording studios; they gave me a job for the summer before I started college. I say "job"—it was only three days a week, unpaid, and really only involved pushing boxes around without trying to get in too many people's way. It was the best time of my life, and I made some lifelong friends that summer. (Maple Studios is owned and run by my mentor, Glyn. This is where my understanding of all things electronic and audible came from, and how I first got introduced to the UK's touring circuit.) After much pestering, I eventually started getting paid—much to my parents' relief.

After that summer, I persuaded Glyn to give me a job in addition to college, and I followed up my practical experience at Chinnery's, the local venue, with what I was learning at college. In all honesty, college was a bit of a waste of time for me—I learned more doing the job than sitting down listening to someone talk about it. You need to bear in mind, though, that the program I was enrolled in was geared more toward recording, and was, I believe, one of only two or three of its kind in the whole country. It was very new, and not many people really knew how to teach the practical side particularly well. I sat in countless classes with a pen and notebook taking notes. And there was me thinking this was a practical job.

My parents, just like all parents, wanted me to be qualified for what I was doing—so I stuck it out, even though I knew it was more about experience. So that was three years of my life. I don't regret any of it, though, because a lot of the knowledge that I gained working in a freezing cold workshop at Maple in the height of winter translated well into the electronics part of the course I was learning at the time.

After college, I went back to work at Maple. My parents always wanted me to get a proper job, but I knew I was doing serious work: There were a grave number of teenage Goths who needed their weekly dose of local ambiguous music, and we were the only ones providing that fix. Slotted in between these local events were some of the best smaller touring acts around, and because I was working, I got to see them. This is where I really began to appreciate the difference between recorded and live music, and these were some of the best years of my life.

One day, we got a phone call from the local theater: They had the Ted Heath Orchestra with Dennis Lotis on vocals coming to town and had no one to mix it. Off I went into the unknown world of theaterland, and some old guys I've never even heard of. But when I told my dad, he knew who they were—and ever since that day he's approved of what I do . . . bless him.

After being in the local venue for a few years, one of the local bands, Engerica, got a small record deal and was going on tour with another band, thisGIRL. They asked if I could come along and mix for them, as I did such a smashing job in Chinnery's. So I went home and said, "Mum, Dad, I'm going on tour"—and off I jolly well went. I slept on mates' floors, in the backs of vans, and once—at TJ's in Newport, South Wales—we were allowed to sleep in the upstairs apartment of the venue. This last one wasn't as glamorous as it seems—we weren't allowed to turn on the gas fire, as it had been condemned. This was a big tour for me: It was on this tour that I first met my now-old chum Pablo (more on him in a minute), survived many breakdowns in the van, experienced countless arguments—and I'm pretty sure it was the same tour where my wallet and the drummer's phone got stolen off a table next to a window in my mate's house in Leeds. Oh, how we laughed!

I met Pablo again when we were out on another tour; he was with the band Kinesis, and I was with another of the Southend-based bands, Smother, which was first on the bill. Winnebago Deal were on in the middle, great little band, just a two-piece, and they were both called Ben. One of the Bens now works at Oxford Academy, and it's always a pleasure to see him. After that tour, I decided to go it alone—so I gave my notice to Glyn, and off I went.

The following year, I was phoning around for some work and called Pablo. He had just started working with a bunch of Welsh Rappers from Newport, South Wales called Goldie Lookin Chain. They needed a monitor engineer, so I jumped at the chance (and also because I didn't have anything else to do at the time). Suddenly, a couple of months later, they had a number one hit in the UK charts: "Guns Don't Kill People, Rappers Do." The next thing I knew, we were touring North America, Japan, and all over Europe, and playing the main stages of some of the best and most prestigious festivals in the UK. I missed my sister's birthday that year because we were doing a show for Channel 4 on a beach in the West Country—but as compensation for not being there, I got the band to say "Happy Birthday" live on national TV while she was watching. I think that did the trick.

So that's me, and the story of how things got going. Most people who mix for a living come from a background of playing instruments and then naturally migrate toward the mixing console. But, as with almost anything, it's really a case of being in the right place at the right time and having the right attitude.

This book has been written on various modes of transport, four different continents, and I-can't-remember-how-many countries over 2009 and 2010, while on tour with the British artist La Roux.

FIGURE I.1
La Roux and Crew (from left to right, Paul Stoney (Backline), Colin Ross (Lighting Designer), Me (FOH), Risteard Cassidy (Monitors) Elly Jackson (Vocals), Mickey O'Brien (Keys), Jess Jackson (Personal Assistant), Mike Norris (Keys), Mark Dempsey (Tour Manager), William Bowerman (Drums).

Thanks, guys, for all your love and support.

BEFORE I BEGINNING...

It's worth recalling where music comes from and why it came about. This might not seem like the most obvious start to a book about live audio engineering, but it's an essential part of understanding what we do.

Music is extremely ancient; ever since humans first started communicating, rhythm and melody have helped tell stories about our past, present, and future. There's something very deep—spiritual even—about music, which I suppose is why it has been around for so many thousands of years. Making music and making musical instruments go hand in hand with this and has been very important to our development as intelligent beings. The early pioneers of musical instruments would pull animal skins over bits of bone or wood and tie various bits of leftover guts to branches.

Music is innate in all of us—it connects us and binds us together, in much the same way that our ancestors would bond around fires. As an audio engineer, you are responsible for making this modern-day fire. You are the musical master of ceremonies. You are translating someone else's emotional and spiritual journey to the masses, and it is a big responsibility.

WHY THIS BOOK

Everything contained within these pages is something that I have come across in my everyday life as live audio engineer, both in house and on tour. These might not be everyday events, but they are certainly things that have expanded my knowledge of the work I love and have found invaluable. The information contained within these pages is what I've found most relevant from my college experience, and what I've learned over the last 15 years of working and playing in the music industry. Some people will have had different experiences than myself—all I can say is that these are problems, solutions, and situations that I come across every day when working for a professional, heavily touring artist.

Some information in the early stages of the book might come across as extremely basic, and even irrelevant; but when we come back across these things later, you'll then understand why we're discussing them. The aim is not to tell you what to do, but rather how to think. The manipulation of sound isn't something that can be read or spoken about. It can be very instinctive, and getting out there and hearing it for yourself is what really counts. Articles and textbooks can tell you how to set up different things, but they never give you the room to discover sound for yourself, which breeds creativity and knowledge, and which is so important in an industry that is as much about technical know-how as it is about emotion.

HOW THE BOOK IS STRUCTURED

I've tried to make this book as relevant as possible by structuring the content in a way that mirrors how live shows work. By reading this book, I hope to help you not only understand the day-to-day life of a touring engineer, but also some of the equipment and thought processes that we go through.

The book is split into two sections: Pre-Show and Show Day. Everything I cover in the Pre-Show section are things you must do or understand before heading out on the road—for example, advancing a show and creating some kind of stage infrastructure.

The Show Day section follows how a show runs in a normal touring scenario, from the moment you turn up to the venue through the process of running a sound check and the pitfalls of putting an audience in a venue. The aim of this section is to talk through events as they happen in real time; for instance, we talk about mics and their placement in Chapter 14. I hope it works in a way that you get the information as you need it, and not have to retain lots of information from previous chapters.

I hope this book comes across as a more real account of live audio, and I hope it is an easy and enjoyable read—because, personally, I can't stand reading textbooks.

SECTION 1
Pre Show

"For the past 32 years, I've done nothing outside the entertainment business. I've had some real highs and some real lows, but I love the work so much that I never once thought of quitting."

—**Meat Loaf**

CHAPTER 1
What is a Live Audio Engineer?

> Job description: If you like semi-darkness, long hours of boredom, long hours of work, no social life, no love life, heavy lifting, getting your white gloves dirty, and a good laugh, this is the job for you.

Audio engineers, also known as sound engineers, come in many different types: TV, radio, film, and live and recorded music, just to name a few. Although these jobs are very different, the people who perform them are all considered to be sound engineers. This holds true for other languages as well: The Germans have different words for jobs such as tone master (*Tonmeister*) and tone technician (*Tonetechniker*), the tone master being a producer and the tone technician someone who operates the equipment.

This book is specifically about live engineers, whose job it is to look after the sound at all types of live events. This can be a high-pressure job, as you only get one chance to get it right. You need to be on the ball, understand when things go wrong, and know where and how to fix them—quickly. In order to help you do this job the best way it can be done, you must have general knowledge of all different aspects of the job.

In a live environment, there are three main types of audio engineers: front of house, monitor, and system technician. In the following sections, we discuss all of these types in more detail.

FRONT OF HOUSE (FOH) ENGINEERS

The front of house (FOH) is where the audience is, and an FOH engineer mixes the audio for that audience. If all goes well, the FOH engineer is the person standing in the middle of the audience next to the lighting guy and surrounded by a barrier and different-colored lights. (The FOH engineer is often mistaken for the DJ, but don't even think about putting a request in.)

FOH engineers work hand in hand with monitor engineers and must have good communication with them. Together, you must follow the band's specifications (see Chapter 5 on Advancing the Show). The FOH engineer also puts the channel list together, thus ensuring that you have all the channels you need to mix the show to your liking. (Remember, though, that there might be some channels that don't need to be heard through the house speakers, such as click channels and ambient mics for in ear monitors (IEMS).) Finally, the FOH engineer also runs soundchecks.

Speaking from personal experience, I have spent some time doing this job, and I always love it; being an FOH engineer gives me the ability to be creative and loud at the same time. However, the mixing can be a challenge. It isn't just a case of pushing up your faders and making sound happen—it's about blending sounds into one another so that you hear a full mix with nothing obscured. This is an enormous responsibility because you essentially have control of another artist's music. Some artists really want to be involved with the mix, while others might just let you get on with what you are doing. Either way the ability to understand what the artist, management, or producer wants, and then the ability to translate that into audio, is important. For example, if someone says "I want it to sound more raw," or "More reverb!" you have to understand what this means and how to do it. We'll get into more detail about this kind of thing in later sections.

MONITOR ENGINEERS

The job of a monitor engineer is probably the most fundamental of all the live engineering jobs. Monitor engineers are responsible for controlling all the sound on stage. *Monitors* are the speakers positioned on stage that allow performers to be able to hear what's going on. They are also referred to as *wedges*, which is the term that most professionals use, or foldback, which is more of an older term that isn't particularly used from day to day. The majority of the work for a monitor engineer is done during soundcheck, making sure that everyone has what he or she needs to hear, and thus perform, well.

You will find the monitor engineer located just off to the side of the stage, preferably on stage left (if room allows it). He or she controls the individual monitor mixes for each of the performers on stage. As a result, it's a good idea for the monitor engineer to put the stage plan together, so that he knows where all his monitors should be and what order the sends for the monitor console need to be on. (We discuss stage plans in more detail in the Stage Plan section in Chapter 5, where we go into more detail about why it's a good idea for the monitor engineer to do it. If there isn't a monitor engineer, this responsibility falls to the FOH engineer.) A monitor engineer might also be in charge of IEMs, or *in ear monitors*. These are similar to headphone buds that can be molded into shape. IEMs can also be *generics*, which are similar to foam earplugs with a headphone attached to one side. There is a real art to mixing IEMs.

In order to be a monitor engineer, the performers must trust your work. This can be a challenge, especially because you may be dealing with big egos. As such, good communication skills are essential for doing a good job. Part of this communication is understanding seemingly random hand signals and gestures.

There is nothing quite like watching an artist wave his arms in the air, point at objects, and nod his head as if some kind of epileptic fit has ensued—all in an attempt to tell you that he requires a little more acoustic guitar in his wedge. One of the basic rules of being a monitor engineer is to pay attention to the performers at all times, looking at them even when they aren't looking at you, and constantly monitoring their individual mixes. Meanwhile, they'll be able to monitor each individual mix with their own wedge—called a *listening wedge*— and their own set of IEMs. Make sure that the listening wedge is exactly the same as the other wedges on stage, with the same amp and the same graphic equalizers. Getting this part of the sound right is essential because performers rely on you to get the best out of what they are doing.

The job of a monitor engineer is probably one of the hardest, but also one of the most rewarding. There isn't much room for creativity, but there is an art form about getting monitor mixes right. When the performers have a great show, the monitor engineer will have a great show.

SYSTEM TECHNICIANS

System technicians, also known as *system techs*, look after the whole PA system. There are normally at least two system techs per PA system—one who looks after the FOH and one who looks after monitors. These engineers are wholly responsible for the entire PA system and usually have a vast knowledge of the equipment they monitor; however, unlike FOH and monitor engineers, they usually do not operate the equipment (unless asked to, or there isn't anyone else to do the job). Although most system techs will be able to mix, their main responsibility is to make sure that all the equipment is working correctly and is properly maintained. The biggest part of this job is to work with the artists' FOH or monitor engineers to get exactly what they need out of the system and equipment.

One type of system technician is also known as an *in-house engineer*. In-house engineers have all the same knowledge as system technicians; the only difference is that they generally work for the venue, whereas system techs generally work for PA companies.

THINGS TO THINK ABOUT

One of the important things for a sound engineer to realize is that it can be quite a social job. You have to learn how to balance this out and realize that you can't simply leave your position to go enjoy the show or visit with friends who are attending. Engineers are not part of the band, even though they may spend a lot of time with them. You must remember that you are the one who has to be in the venue before the band members, get things set up before them, and stay behind after they have gone. Just remember: You are there for a reason, and that reason is to mix. You are being paid to do that and nothing else.

Some engineers I know have next to no negotiation skills; their attitude is "It's either my way or the highway." Try not to adopt this attitude when working—it's very important to be able to adapt to your surroundings. You are part of a team that makes everything work, and it's everyone's job to make sure the show happens and that the audience has a great time.

This industry is still very young and, as such, is constantly changing. Just remember that if you want to make a career out of it, you must be professional, responsible, and courteous at all times. In addition, here's a little tip for when you're on the road: You never know how someone else is dealing with being away from loved ones, so you should always give people the benefit of the doubt.

Above all, remember this: You are only as good as your last gig.

HOME LIFE

Your home life is one of the hardest aspects to navigate in this type of career. Some people are built for travel, whereas others are made to stay in one place. Some engineers get into this job because they are very attracted to the idea of being able to see the world, even if only from the back seat of a taxi, or looking out at a cityscape through a window in a departure lounge. However, having a stable life at home is key not only to your own sanity, but also to the sanity of the people around you. Being away from home can put a strain on even the most solid of relationships, but the key ingredient for any type of relationship is communication. With this type of job, having a family that understands who you are, what you do, and why you do it is extremely important.

One of the difficulties of being a live engineer is getting outsiders to understand what the job is like. Many people have incorrect preconceptions, especially due to the kinds of stories you hear about the early days of rock and roll. These days, though, things are very different; usually you get straight on the bus after a gig and head straight out of town. Going on tour is about making money, which means you are always on the move.

Having troubles at home while you are away can lead to all sorts of problems. It nearly always affects your work because your mind is constantly taken away from the job at hand. As such, it can also affect the people you are working with. Chemistry is crucial on the road, and a breakdown in trust and communication can be disastrous for the whole operation. Just remember, it is one thing to talk about troubles, and another to take your troubles out on other people. Take time for yourself. Everyone is in the same boat and will understand if you don't want to be part of group activities outside work time.

CHAPTER 2

Audio Engineering Basics

If you want to be a sound engineer, you must have a good understanding of all the elements that affect the job. This section explains these elements.

THE EAR

Your ears are one of the only senses that aren't ever turned off. You can close your eyes and stop touching things, but your brain is always processing audio. According to certain studies, audio frequencies affect brainwaves; for example, the complex patterns of Beethoven's music stimulate the brain and thus improve thought processes, helping you to retain more information. Although it's hard to vouch for this personally, many people will tell you that they have strong reactions to music they hear—perhaps even a built-in passion. What is clear is that music does affect the way we think. For example, in my personal experience, the simpler the music being listened to, the easier it is to relax, whereas the more intricate the music is, the more stimulated I feel.

Sound can be an incredibly powerful sense, and not only for animals who "see" using sonar, such as dolphins and bats. In Dorset, United Kingdom, there is a 7-year-old blind boy who navigates using a series of clicks. The technique is called *echolocation* and was developed in California; it is based on the Doppler effect, which is the principle stating that when an object is moving away from you, it creates a lower pitch, and when it is moving closer to you, it creates a higher pitch. For example, when a police car goes speeding past, you'll hear the frequency of the siren change. Using this knowledge, and with much practice, practitioners of echolocation can determine the height, width, and location of specific objects; in some cases, they can even guess their density. Objects that are closer, larger, and simpler are easier to perceive. A technique like this shows us how we can use and harness the power of frequencies, and how important it is to look after our hearing. It also says a lot about what can be achieved using audio and how important it is for everything we do.

Do We All Hear the Same?

The simple answer to this question is no. It's not that our ears work differently; rather, due to a number of factors such as ear size and damage, the frequency ranges each person hears vary slightly. If you are right-handed and have been a drummer all your life, for example, you might notice that, in your left ear, you may not hear 1.5 kHz very well. This is likely because of your snare drum. Your perception of sound also depends on how you listen. Have you ever heard an album and thought, for example, that the cymbals sounded great—but then learned that others disagree? This is because listening is, to some extent, dependent on experience. You can train yourself to listen to different things, either individually or collectively. As an audio engineer, you need to separate and sort sounds in your head, and you also need to determine from which direction they are coming.

How Do Ears Work?

The ear consists of three parts: the outer ear, the middle ear, and the inner ear. When you first hear sound, the pinna (which is the name for the outer ear) captures the sound and funnels it into the ear canal; then, once it travels through the ear canal, the sound reaches the eardrum, which is the border between the outer and middle ear. The middle ear consists of three bones, collectively

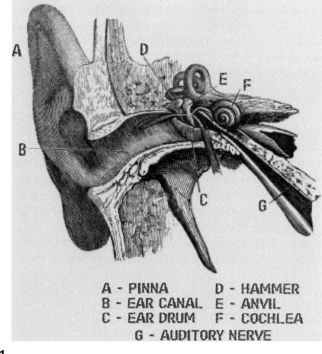

A - PINNA   D - HAMMER
B - EAR CANAL E - ANVIL
C - EAR DRUM   F - COCHLEA
     G - AUDITORY NERVE

FIGURE 2.1
The ear.

called the *ossicles*, but individually named the hammer, the anvil, and the stirrup. As the sound makes the eardrum vibrate, the vibrations from the drum get transferred to the hammer, the hammer vibrates the anvil, and then the anvil transfers the vibrations to the stirrup.

The cochlea, which looks like an ammonite, is the inner ear. The inside of the cochlea is full of fluid and contains tiny hair cells. As the liquid reacts to the vibrations from the middle ear, these tiny little hairs move. The movements of the hair cells are turned into electrical signals and picked up by the hearing (auditory) nerve, which in turn sends the signals to the brain. The brain then translates the electrical pulses into what we perceive as sound.

Ears and Frequencies

The frequency response of the ear is not by any means flat. Our ears are trained to listen to voices, so our hearing is at its best between 1 kHz and 5 kHz, with some variation depending on the individual. These are the frequencies that allow us to hear clearly; if you were unable to hear these frequencies, you might be able to tell the difference between voices, but you wouldn't be able to understand what they were talking about.

Perceiving Loudness

The ear is a very sensitive organ. The range of power we can perceive is vast and is measured in watts per square meter (W/m^2). The smallest amount of watts we can hear is about .000000000001 W/m^2, and the largest amount of watts we can physically stand is about 1 W/m^2. These ranges are quite extreme, and there are a lot of numbers in between them, which should give you an idea of the vast range of our perceived loudness. That being said, here are two things to keep in mind. First, if you went around listening to everything at 1 W/m^2, you would promptly go deaf; and, second, you would need to be a child who grew up in the desert and never heard anything louder than a fly buzzing in order to have hearing that detects .000000000001 W/m^2.

Experts say that 85–90 dB SPL (Sound Pressure Level) is a safe hearing level. However, when you consider that the average noise coming out of a lawnmower is 90 dB SPL, it's unrealistic to expect that all music should stay within this range. In fact, 90 dB SPL is the point at which you start feeling music, the point at which you can feel the vibrations in your feet. Most concert music is around 100–110 dB SPL, though this depends on whether the venue has any restrictions, as well as on how it is being measured (please refer to the Decibels section in Chapter 3). Anything under 100 dB SPL, and you'll probably have the artist's management telling you to crank up the volume.

Although you start to experience physical pain at 140 dB SPL, in reality, your ears will start being dramatically affected around 110 dB SPL, and you'll have a problem, even ringing, in your ears for days. If you ever hear 140 dB SPL, you may not hear anything ever again.

Reflections

We hear reflections in sound all the time, and it's these reflections that make a sound seem natural. Every time a sound is made, the waveform is sent out in every single direction. It'll then come into contact with various objects, kitchen sink, a wall, anything that is near it. Nearly all objects reflect sound, and it's these natural reflections that we hear along with the source sound that make it sound natural; if you were to hear a sound without any reflections, it would sound very alien to you.

Hearing Localization

As you know, sound levels and frequencies help us determine what kind of sound we are listening to, and the difference in sound at the two ears helps us tell from where a sound is coming. Your brain can detect the very smallest of delays in a sound reaching both ears, which is what allows you to determine its direction. Hearing localization is very important—stepping into a war zone is hazardous in the best of times, but if you weren't able to hear where sounds were coming from, it would be suicide. Of course, hearing localization is also helpful in more everyday scenarios.

We enjoy listening to music in stereo. It gives us an audio image of space within the sounds. We can pick the sounds we want to listen to far more easily. Why do we place two speakers apart from one another rather than just one in the middle? After all, just placing one single speaker in the middle of the room would serve its purpose and use up a lot less space.

We have two planes of hearing; a plane is a flat surface. One of these planes is oriented horizontally, and the other is oriented vertically. The physical placement of our ears on the side of our heads gives us an extremely high-definition hearing range along the horizontal plane. If a sound is directly in front of you, the sound reaches both your ears at the same time; this is how your brain identifies the location of that sound. However, when the sound is coming from the left or right, it takes longer for the sound to reach the more distant ear. This tells your brain whether the sound is to the left or right of where you are standing. Along this horizontal we can hear a difference of only 1 or 2 degrees, or in terms of the time it takes to a sound to arrive between the two ears, about 13 microseconds. To stress the importance of this fact, and why our hearing is so defined, think about a film. The frame rate, which is how many frames a second it takes for us to see a complete moving image, is between 24 and 28 fps (frames per second), compared to an audio frame rate (the amount of audio information we can process in a second) of approximately 56,000 fps for audio. It's quite a lot, but still audio seems to come second to visuals.

The way localization in the vertical plane works is a little more complex, but it still uses the same principle. The shape of your pinna causes reflections, which in turn create small delays. It is the difference between the direct path and the reflected path that helps us work out if a sound is above or below us.

HEARING LOSS

As should be expected, hearing loss is very common in an industry where high noise levels are accepted. There are several different types of hearing loss, which we discuss in this section.

Types of Hearing Loss

CONDUCTIVE

Conductive hearing loss is caused when something stops the movement of sound traveling from your outer ear to your inner ear. This can be caused by a buildup of earwax blocking the outer ear or by pierced eardrums, which are often the result of untreated ear infections, head injuries, poking something down your ear, or a collection of fluid in the middle ear (known as *glue ear*).

SENSORINEURAL

Sensorineural hearing loss is caused by actual damage to any of the components between the inner ear and the brain; it harms the hair cells inside the cochlea, a process known as *acoustic trauma*. Sensorineural hearing loss also affects the intensity of sound, making it difficult to hear complex sounds, especially in noisy environments. Sensorineural hearing loss is probably the most likely type of hearing loss in loud environments, and is caused not only by volume, but also by the length of time you are in the loud environment. However, it can also happen naturally, due to your age (a process called *presbyacusis*), or, alternatively, it can be caused by infections, certain cancer treatments, or other medications.

In general, you should always keep your hearing in mind. If you are having any form of treatment for hearing loss, remember this when you are mixing, and take it into account.

Prevention

CERUMEN

Also known as earwax, cerumen is your ears' natural defense against loud sounds. If you are frequently in loud environments, you'll find that a buildup of wax slowly appears. A buildup of too much wax can cause a buildup of pressure and be quite painful, but cleaning them will help. Infections also can cause earwax, so be aware of this possibility if you are mixing when you have a cold. Earwax is only a natural defense and can't be used as any form of substitute for proper hearing protection.

EARPLUGS

The only effective way to prevent any form of hearing damage or loss is to use earplugs whenever you are in a loud environment. Obviously we are working in loud environments, but you can't wear earplugs while you are mixing because then you won't be able to hear what you are doing. The best thing to do is to minimize your exposure to excessive sound levels by using earplugs when you aren't mixing in those environments such as when you are setting up and the support band is playing—and I know this isn't the coolest thing to do, but also when you are going out for a night clubbing. When you *are* wearing earplugs, the best ones are the big foam plugs that cut out everything. Some engineers actually use foam earplugs after setting up each song's mix, taking them in and out as needed. You can get impressions made of your ear and molds made from these impressions, which reduce sound levels. Though they are pricey, they are worth the money. Or if you want to look super cool, you could get some ear defenders like the ones worn by the guys doing all the drilling outside my house.

EAR HEALTH

Obviously, it's important to keep your ears in a healthy condition. The first rule of good ear health is to never put anything into your ear that is smaller than the size of your elbow. Earwax does serve a positive purpose and should not be totally removed; it is there to help filter out dust or other alien objects from your ears, as well as moisten your ear canal. Without earwax, your ear canal would be dry and itchy and rather unpleasant. If you wash your hair on a regular basis, this is enough to keep your ears nice and clean, but getting them cleaned out by a medical professional every few years won't harm them either. And by all means, you can of course clean the outside of your ear. Some people overproduce earwax, in which case doctors can prescribe some medication to help clear it up. To repeat: Don't put anything down your ear.

HOW DOES SOUND TRAVEL?

Sound is the vibration of any object. Because we live in an environment that is mainly air, then, sound is the vibration of air. When vibrated, air molecules hit other air molecules and eventually vibrate in your eardrum. Soundwaves can also travel through any material; as long as it has molecules to pass through, it will travel from one molecule to the next

Obviously, this is a simplification of the process, but it is enough for our purposes.

Speed of Sound

The speed of sound is commonly misquoted as 340.29 meters per second at sea level. However, the speed of sound isn't a fixed speed; it all depends on temperature, humidity, and pressure. If the air pressure, temperature, and humidity are nonvariable, the speed of sound is the same at sea level as it is at the top of a mountain. However, if any of those parameters changes, so does the speed of sound. Similarly, sound travels faster through objects that are denser than air, such as water or steel. Because the particles that make up these things have a higher density, the information transferred between molecules is quicker. So if you were to shout through a tunnel while banging a metal pipe at the same time, your buddy who is a couple of miles away would hear the banging pipe first and then a short while later hear your voice.

Changes in atmospherics have a lot to do with the speed of sound. If the outside air temperature is cold, the sound will travel more slowly because the molecules in the air have contracted and they have restricted movement. But when you warm the air up they become less restricted and have more space to move. So think about this when you are looking at a PA system in a stadium with one-half in direct sunlight and the other half in the shade. Another important consideration is the humidity. Because water is denser than air, the humidity in air transfers sound more quickly than less humid air. When air becomes hot and humid, there are more water molecules in the air and the air molecules have more space to move, so the speed of sound is increased even more.

WAVEFORMS

A waveform is a graph of the amplitude versus the time of a sound. There are infinite amounts of waveforms, all of which have their own characteristics and sounds, but we can only hear a very small section of them. Waveforms represent every sound we listen to, and they follow a complete cycle, starting at zero volts, rising to a positive peak, returning through zero volts to a negative peak, then returning to zero to complete the waveform.

As live audio engineers, you will likely deal only with pure tone waveforms, or, as they are more commonly known, sine waves. Even so, it's a good idea to have an understanding of other types of waveforms; after all, you never know when you might need to look at a synthesizer.

You can manipulate a waveform very easily. For example, when distortion is applied to a guitar, the signal voltage can only go so high. This causes the top of the wave to be flattened, making it look more like a square wave. Because some of the source signal is changed, the effect is distortion.

Types of Waveforms

In this section, we discuss the various types of waveforms.

SINE WAVES

A sine wave is the purest and simplest waveform of all; all other waveforms are sums of sine waves. A sine wave is perfectly symmetrical, smooth, and repetitive, and its oscillation keeps its shape through its entire cycle. The peak-to-peak values stay the same. You'll notice when listening to a sine wave that you'll just hear one pure frequency; it'll be smooth and constant in volume.

Adding together different sine waves gives you different types of waveforms. These sine waves can have different phases, frequencies, and/or amplitudes.

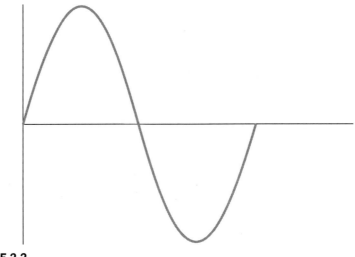

FIGURE 2.2
One full cycle of a sine wave.

SAWTOOTH WAVEFORMS

There are two types of sawtooth waveform. The first type, which is just called a sawtooth wave, increases rapidly at the front and then decays. The second type is referred to as either an *inverse* or a *reverse* sawtooth because it is the inverse of the first. In both types, the wave looks like a tooth on a saw—hence the name—and in both types the waveform sounds identical. Sawtooth waves contain both odd and even harmonics, and thus have very clear and harsh sounds. When you listen to them, you'll hear a kind of dissonance that might not sit right in your ear. They are frequently used in synthesizers because they are able to re-create analog instruments such as violins.

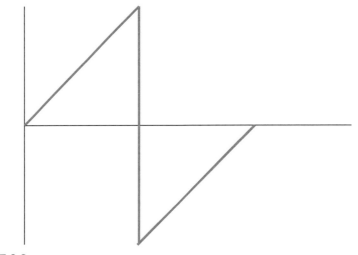

FIGURE 2.3
One full cycle of a sawtooth wave.

SQUARE WAVEFORMS

Square waves are also used in synthesizers, but they come from simple logic circuits and are basically simple binary on and off commands. They contain a great number of only odd harmonics, and because of the number of odd harmonics they have, they sound harsh and processed. Their sound can often be described as hollow and very rigid.

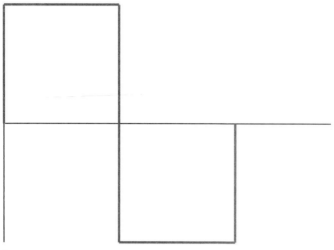

FIGURE 2.4
A full cycle of a square wave.

TRIANGLE WAVEFORMS

A triangle waveform only contains odd harmonics, but isn't made up of as many different types of harmonics as a square wave. When you look at it, it resembles a sine wave, only it is more pyramid shaped. A triangle wave is fairly

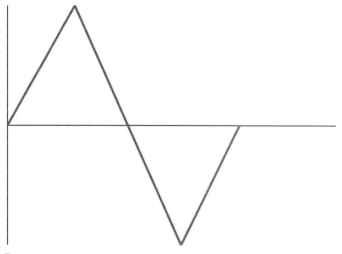

FIGURE 2.5
A full cycle of a triangle waveform.

full sounding. It's not too harsh, and it doesn't grate on your ears as much as a sawtooth or square wave does.

Waveform Power

Every signal has a power associated with it, but the power differs for two different types of waveforms with the same peak-to-peak value. For example, let's consider a square wave with a peak-to-peak value of $+/-5$ dB, and say it is 100% power. Because a square wave switches between $+$ and $-$, there is practically no rise time (i.e., the time the waveform takes to get from peak to peak); in other words, the waveform stays pretty much full power all the time. A sine wave, on the other hand, has half the power in the same peak-to-peak range because it spends half its time rising and half its time lowering. Both triangle and sawtooth waves have a third of the power of a square wave.

Understand the different power carried in different types of waveform is the difference between blowing an amplifier sky high, or keeping it in its nominal working conditions, because the signal strength of a square wave is twice as much as a sine wave that the amplifier accepts.

To summarize:

- The power from the square wave is 100%.
- The power from the sine wave is 50%.
- The power from triangle and sawtooth waves is 33%.

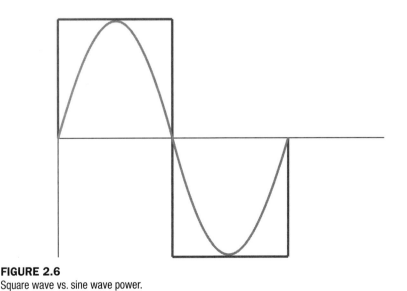

FIGURE 2.6
Square wave vs. sine wave power.

HARMONICS

A harmonic is a multiple frequency of a fundamental frequency.

Complex sounds are made up of many different harmonics. The sine wave, being the purest, doesn't have any harmonics, but once you start to add more and more harmonics you can create lots of different sounds. For example, when we play middle A (or Concert A) on the piano, the lowest (fundamental) frequency the string vibrates at is 440 Hz, but when you listen to the sound you can quite clearly hear that it is made up of more than just frequency vibrating at 440 times a second. The other frequencies are the harmonics of middle A.

There are two different types of harmonics: odd and even. These are just like odd and even numbers. The main frequency is called the *fundamental*, and this is the lowest frequency in a complex waveform. It usually determines the pitch of a note. For example, let's assume that 100 Hz is our fundamental frequency and our first harmonic. If we multiply this number by 2, we get the second harmonic: 200 Hz. Because we multiplied the number by an even number, we call this an even harmonic. If we multiply this number by 3, we get our third harmonic: 300 Hz. Because this number is multiplied by 3, which is an odd number, this is an odd harmonic. In other words, if you multiply by an even number, you get an even harmonic; if you multiply by an odd number, it is an odd harmonic.

Looking back at our waveforms, we see that a sine wave is just one pure tone and does not contain any harmonics. By adding together different sine waves that are harmonics of the original fundamental frequency, we get our different types of waveforms. By changing the phase and the amplitude of these different sine waves, we change the overall sound.

Harmonics that make up complex sounds can be broken down into different categories, two of which are the following:

- *Asymmetric:* When we have asymmetric harmonics, the waveform can contain either or both odd and even harmonics. The asymmetrical nature of these waveforms means that the harmonics aren't equally distributed on either the positive or negative side of zero.
- *Subharmonic:* When looking at our fundamental frequency of 100 Hz, if we divide that number by 2, we get 50 Hz. This is a subharmonic frequency of our fundamental. The series runs the same as our harmonics, except we divide rather than multiply.

Overtones

An overtone is exactly the same as a harmonic, except it is labeled slightly differently. The first harmonic is the fundamental frequency. The first overtone is the second harmonic.

TERMS AND DEFINITIONS

In this section, we discuss some of the most important sound-related terms and definitions.

Hertz, Cycle, and Frequency

Hertz, cycles, and frequencies are all closely related. A cycle is the complete oscillation of a waveform—that is, the time it takes for a signal to go from 0 volts, through its top peak and bottom peak, and return to 0 volts again.

Frequency is the number of cycles in one second, and these are written as *Hertz* (Hz). Within the frequency range that we hear, these frequencies are called audio frequencies (or AF, as it is sometimes abbreviated on equipment).

When two frequencies start getting close to each other, you can hear them start to pulsate, or make a beat. You may have noticed this when a guitar is being tuned: When the guitarist or his tech is tuning one string against another, you can hear a pulse in the notes as the notes get close to each other. When you have two frequencies—let's say 290 Hz and 300 Hz—that are very similar in frequency, you can easily mistake them for the same frequency when they are played on their own. However, when they are played together, you can hear the frequencies pulsate. You are actually hearing the two strings pulsate at a rate of 10 beats per second. As the two frequencies get closer and closer together, you will hear the beat slow down into one long note. Listening to the pulses within the frequencies is a great way of determining whether something is in tune.

When we talk about frequencies, you will see them abbreviated as Hz. You will also see following:

- kHz: Kilohertz. This is the equivalent of 1,000 Hertz.
- MHz: Megahertz. This is the equivalent of 1,000,000 Hertz.
- GHz: Gigahertz. This is the equivalent of 1,000,000,000 Hertz.

Both megahertz and gigahertz are very rarely used in audio as they are way out of our hearing range, but they are used in the world of RF (Radio Frequencies). We use gigahertz more often to describe the frequency that a mic or an in-ear monitor pack is sending or receiving.

Wavelength

A *wavelength* is the distance between two peaks of a wave of sound—or any other type of wave, for that matter.

Amplitude and Loudness

Amplitude and loudness go hand in hand. Looking at a waveform, you will see that one full oscillation has a maximum peak level across the entire waveform. If you have a waveform that only reaches +4 dB and −4 dB, that waveform has less amplitude than a waveform that reaches +12 dB and −12 dB. The more amplitude you have in a waveform, the louder the signal.

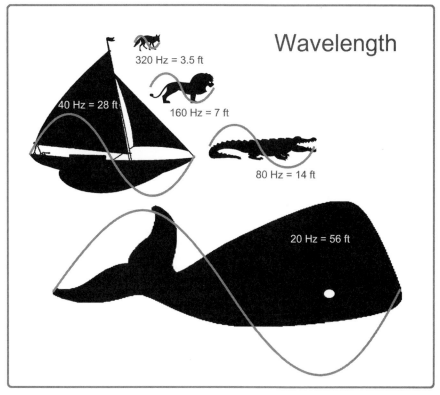

Wavelength

320 Hz = 3.5 ft

40 Hz = 28 ft

160 Hz = 7 ft

80 Hz = 14 ft

20 Hz = 56 ft

FIGURE 2.7
Each sound wave in air has a physical length.

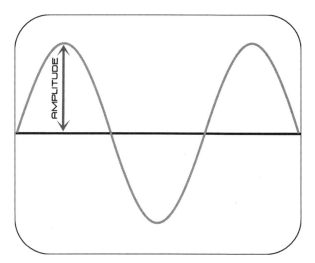

AMPLITUDE

FIGURE 2.8
The height of the waveform is its amplitude. It correlates with how loud the signal is.

Phase and Polarity

Throughout this book there will be terms that are used every day in the live audio world that might not actually be 100% scientifically correct. I've tried to point them out as best I can, and here is one such example. You'll hear this a lot: the word "phase." When we say something is out of phase, or speak about hitting the phase switch, we are actually talking about the opposite polarity. The angle of phase varies from 0 degrees right through to 360 degrees, so when something is actually out of phase it's the variant across all 360 degrees, whereas when something is 180 degrees (this is the point of complete cancellation) out of phase, it's actually the opposite polarity.

Phase is probably the most important aspect of sound for us as audio engineers. It's the fundamental building block for Equalization (EQ), and it also glues together all the sounds we hear. It is easily misunderstood, but when properly understood, it can be very beneficial.

When we look at the cycle of any waveform, as we have been doing, we see the waveform cycle through its peak to peak; it is also cycling through 360 degrees of phase, where 360 degrees equals one complete cycle.

A waveform isn't just the 2D image we see on pages of physics books all over the world. Waveforms are actually complete cylinders, and if you look at the cylinder head on you'll see nothing but a complete circle. This circle can then be broken up into degrees just like a compass.

If we pull out our phase compass and point it due north, our phase will be 0°; the opposite of that would be due south, which would give us a phase (or

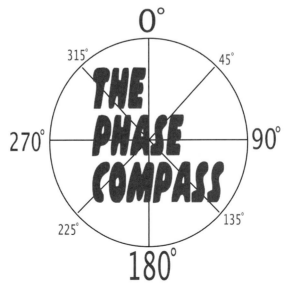

FIGURE 2.9
360 Degrees of phase.

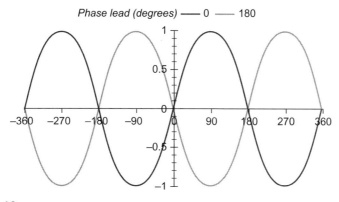

FIGURE 2.10
Out of phase.

polarity, hence the north and, soon, south metaphor) of 180 degrees. When we have two waveforms carrying the same information, one of them is pointing north and the other is pointing south—they are exactly opposite; and when we talk about waveforms being opposite we talk about them canceling each other out. This is where a signal being out of phase comes in.

When two identical signals are 180 degrees out of phase (opposite polarity), they cancel out so we don't hear anything. To give you an example of how this can be a problem in everyday live-sound scenarios and also studio scenarios, consider the case of a snare drum. We normally mic up a snare with two mics, which are referred to as top and bottom, respectively. When a snare drum is being hit, the skin of the drum moves in the direction in which the drum is being hit. The mic on the top of the snare drum picks up the waveform moving away from it at the instant the drum is hit; however, on the bottom, the bottom skin moves *toward* the mic, so the mic picks up the waveform moving toward it at the instant the drum is hit. When the signals blend at the mixing console, we have two opposite waveforms. Because the sound from the top and the bottom skins oppose each other, the low frequencies in the snare drum cancel out, giving a thin sound.

Now we know that when a signal has a phase shift of 180 degrees relative to an identical signal, it is completely canceled out, but there are different degrees of phase shift all the way through our phase compass. As the signal moves away from 0 degrees and the phase shift gets closer to 180 degrees, the cancellation become more severe. Once it reaches 180, it is completely canceled out, and as it moves away from 180 degrees through to 360 degrees the cancellation gradually becomes partial, until it reaches the 360th degree; then we are back at zero again, and our signals are perfectly in phase again.

Always keep in mind that when something doesn't sound right, phase cancellation might be the problem.

FIGURE 2.11
The phase of a snare drum.

Pitch

Pitch is the perception of the frequency of a sound. This might not be the fundamental frequency, but sometimes it is the second harmonic that contributes to a sense of pitch.

Transient and Timbre

Transient information is one of the most important factors in any waveform. A transient is a rapidly changing signal, such as the attacks at the very beginning of the waveform.

Timbre is the perceived tonal differences between pitch, loudness, and length. It's the color of what we are listening to, and along with transient information it helps us to define the difference between a violin and a guitar, for example. The timbre of each musical instrument is caused because of the amount of differences between the complex waveforms that are produced. If a violin and a guitar play the same note, at the same level, for the same amount of time, we still know the difference between them because they produce different harmonics.

Transients can be measured in rise time. Rise time is at the start of the sound envelope and is the time taken for a signal to go from minimum level to maximum level. Let's look at something with a very short attack time, a snare drum. The time it takes a snare to go from no sound at all to its maximum peak level is extremely short. The sound that defines a snare drum is carried in that very small amount of time. If you listen to a snare closely, all the snap and the punch come at the same time, at the very beginning of the sound.

Reproducing transients accurately is essential in order to clearly reproduce the defining consonants that make for good vocal intelligibility. They make the audio exciting, and they carry most of the definition of the sound. As such, they are as important, if not more important, than the frequency response itself. Without the leading edge or attack of each note, we lose the impact that the attack carries. (Later in the book, we'll explain how transients affect your mix and how you can use or not use them creatively.)

Transients are extremely important in the manufacture of speakers and PA systems because speaker components also have a rise time. If you have a source signal with a rise time that is less than the time taken for the speaker to reproduce it, you will be losing signal definition. Therefore, in the manufacturing of speakers and PA systems, it is extremely important that the transient information be as accurate as possible.

Feedback

Feedback is the squealing or rumbling tone you hear when sound from a loudspeaker goes back into the microphone and is reamplified. This causes a never-ending loop of audio, usually at a specific frequency or frequencies. This can be very dangerous: It can harm loudspeakers, and it can cause hearing damage.

Although feedback within a PA system is bad, many rock bands use feedback between their guitars and amps as part of their sound. Many of the wailing guitar tones that we grew up with and love are all caused by this effect.

Envelope

Envelope is a line connecting the waveform peaks of a single note. It is the change in amplitude over time of one note. An envelope has four parts: attack, decay, hold, and release. (We use these same terms for other things, such as reverb units, compressors, gates, etc.; for example, the attack of a waveform means the same as the attack of a compressor.)

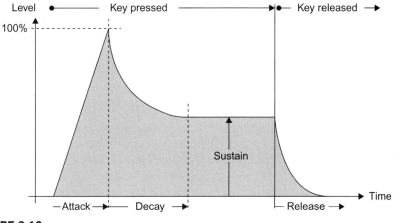

FIGURE 2.12
The envelope of a waveform.

ATTACK
Attack is the first stage of any waveform, and it carries the transient information we discussed earlier. It is the part of the waveform where the note rises from zero volume to its maximum volume.

DECAY
When the attack peak is reached, the decay comes immediately afterward. The signal will decay until it reaches a constant level, either falling into nothing or reaching a sustained level. However, some notes do not decay. If the sustained level is at the same level as the attack peak, you won't get a reduction in sound level.

SUSTAIN AND HOLD
Sustain is the period after the decay and before the release; the note can either remain at the same amplitude, decrease, or increase. *Hold* means the same thing; you will probably see this more on outboard units, such as gates.

RELEASE
Release is the length of time it takes for the signal to drop to zero level.

DECIBELS

The exact definition of a decibel (dB) consistently puzzles people in the sound industry. It is a difficult concept to define; in fact, most audio engineers probably don't quite know how to define it. Should you ask an audio engineer, you may get a convoluted answer that doesn't make sense when put in a different context. Despite this, you will come across dBs all the time; they are used to help monitor the level of a signal and how loud a sound is in a particular environment. They also serve as a level reference point.

The fact of the matter is that decibels are so easy to use as a reference number that there are so many different types used, and so the original use and meaning have been somewhat watered down. With a little thought and coupled with general ignorance, we can pretty much change the way decibels work from power to pressure and voltage. And this is where all of the confusion comes from (I even get confused myself), trying to remember when to increase by 3 dB to double the power or whether we have to double 6 dB to double the power. But what do we really need to know about decibels? Well, not a lot. As live audio engineers, we rarely need to know which type of decibel we are using, and even more rare than that we need to know how to work them out. But the reason we use them is that it's easier to say 120 dB than to say the dynamic range is 20 to .00002.

With all these different types of decibel, we need to know what they all mean. We do so by adding an abbreviation after the dB part: for instance, *dB SPL*. This helps us to know what figures we are looking at because 0 dB SPL is very different from 0 dBV. Another problem is that most of the time when you look at a decibel there is no abbreviation after it, so then you wouldn't be mistaken for thinking that they were all the same thing.

As live audio engineers we will most of the time be dealing with sound pressure levels (SPL), but we'll also come into contact with two other types of decibel:

- dBV (Volts), which is what your mixing console uses
- dB FS (Full Scale), which is the scale we use for digital

There are many other types of decibel, which are fairly irrelevant to live audio engineers.

What do we really need to know?

- A decibel is a *ratio* between *two numbers*. These numbers can be power, voltage, or pressure, but unlike pounds or miles, it is not a set unit, and therefore it is hard to quantify.
- All decibels are relative, and because they are a ratio we need a reference point, which is 0 dB.
- You need to know that on a mixing desk 0 dB is ok to hit.
- When we are talking about the volume, we hear as a sound pressure on our eardrum that 0 dB is next to no sound at all, and 110 dB is bloody loud.

Continuous exposure to anything above 85 dB can cause hearing damage, and 95 dB is the point at which we start to feel the music.
- On some digital consoles the visual scale you see is dB FS (full scale), which is a digital scale where 0 dB is the maximum level you can put into the console before your signal becomes distorted. You can tell if you are working in dB FS because 0 dB will be at the top of the meter on the channel.

Let's explore these three types of decibel a little further.

dB SPL: Sound Pressure Level

db SPL is a measurement of the pressure that sound has at a point, which correlates with how loud the sound is.

Regardless of whether you have a 500 watt PA in a small space or a 50,000-watt PA system in an open field, you can still achieve 100 dB SPL at the mix position—but you will only be able to achieve that kind of dB level with the 500-watt PA in an open field by moving closer and closer to the PA system.

The measured dB SPL at any point in a room depends on:

1. The sound level that the source (speaker) is putting out at 1 meter.
2. The distance the measurement device (or listener's ears) is from the source.

dB V: The Voltage Ratio

When we see 0 dB written on a mixing console, it'll usually be referring to dBV. Because analog mixers use varying voltages to change volume, EQ, gain, and pretty much everything else they use, 1 volt (V) was decided to be used to represent 0 dB, and then the variation in voltage above and below 1 volt gives us our +/−dB.

This is really important when looking at digital systems because you need to understand that 0 dB is the maximum level, rather than the best signal level. If you look at your meter bridge you may notice that at the very top is written 0 dB, this is the absolute maximum level that channel can take, and for those of us who were taught to have our gain structure at 0 dB on old analog consoles will be in for a big shock.

dBFS: The Digital Decibel

The FS in dB FS stands for *full scale*. This is really important when looking at digital systems because you need to understand that 0 dB is the maximum level, rather than the best signal level. If you look at your meter bridge you may notice that at the very top is written 0 dB, this is the absolute maximum level that channel can take, and for those of us who were taught to have our gain structure at 0 dB on old analog consoles will be in for a big shock.

Let me elaborate, 0 dB in full scale is the maximum digital level before clipping occurs. In Chapter 9, we'll look at *bit depth* more closely, but for now we'll consider why bit depth is related to a decibel. When we have digital audio that is

8-bit, 16-bit, or even 24-bit we're referring to the highest possible level that can be produced by that piece of digital equipment.

In 8-bit audio, the maximum dynamic range for our signal is 48 dB. And, as with all things digital, the higher the bit depth, the more you get out of it. If you double the bit depth to 16 bit, your maximum range becomes 96 dB; for a 32-bit, you have a range of 192 dB.

Because we know what the maximum decibel limit is we can use that as our reference point. So 0 dB FS becomes the maximum level we can have before the signal enters digital distortion (which doesn't sound too good), and because our signal is referenced to that all our readings will be in negative numbers.

With new digital consoles, you will see that 0 dB is just above the middle of the fader like it is on analog consoles, and you can still move your gain knob from negative to positive. All that is happening here is that the consoles themselves are using the dB FS scale, but the numbers you are reading have been translated into the old format, so you can more easily relate to what the desk is doing.

Definition

If you really want to know the actual definition of a decibel here it is:

$$10 * \log(P_1/P_0)$$

+4 dB u VS −10 dB V: PROFESSIONAL GEAR VS. CONSUMER GEAR

From time to time, you may see the value +4 dB u and −10 dBV written on various pieces of equipment. These are understood as the professional (+4 dB u) and consumer (−10 dBV) operating levels; in other words, they are the nominal signal levels that pro gear accepts and the nominal signal levels that consumer gear accepts. dB u and dBV are both a way of comparing voltages; it just so happens that the industry standard +4 dB u came about before dBV existed. There is a difference between dB u and dBV, but they both measure volts. The difference between +4 dB u and −10 dBV is not actually 14 dB; it's 11.79 dB. So if you connect a +4 dB u signal to a −10 dBV input, you need to attenuate (decrease) the signal level by 11.79 dB to prevent distortion in the −10 dBV (consumer) device.

You come into contact with electronics every time you use a mixing console or effect unit, speakers, amplifiers, and the like. With the birth of consoles and processors, electronics in this new digital age might seem very complicated; however, the initial principles are the same as always.

In this chapter, I'm going to provide a quick refresher on the basics of electronics and then go into a little more detail about ohms and power.

THE ELECTRONIC OLYMPICS

The most critical part of electronics is understanding how a circuit works. This will help you understand the process of patching up mic cables in a stage box, connecting speakers together, and, most importantly, re-patching.

The Circuit

An electronic circuit is like a 400-meter running track. It has a start and it has a finish, both in the same place. If you think of a battery as the start and the finish, you must put a track out to connect the two poles (your positive and negative terminals) together. On this track, we can put any number of devices that require power—for example, a light bulb. The power starts from the battery − terminal, it goes around to the light bulb and through the light bulb, and then it finishes back at the battery + terminal.

The Athletes

In this analogy, the athletes are the electrons held inside the power source. It is the electrons passing though the various different components of a circuit that creates power in these components. Electrons are negatively charged, and, as with anything, positively and negatively charged objects are attracted to each other. Therefore, the flow of electrons in a circuit always goes from negative to positive. The light bulb has two terminals on either side. One is connected to the negative terminal on a power supply, and the other side is connected to the positive terminal on the power supply. This means that the circuit is complete and that our athletes are free to run around our track from the start (negative terminal on the power supply) to the finish line (the positive terminal on the power supply). While they are going round the circuit, they go through the light bulb, causing the light bulb to light up.

Conductor

A conductor is anything that electrons can pass through to get from one end to the other. In our analogy, this is our running track. The conductor can be in the shape of a copper trace on a printed circuit board (PCB) or copper wires connecting the two terminals of a speaker to an amplifier.

PCB

PCB stands for *printed circuit board*. Most of the equipment you will use has a printed circuit in it. Although you don't need to know exactly what each component does on the PCB, you should understand the basics so that you can recognize the telltale signs of damage on a PCB.

Components can burn out and create a black mark. If you see one of these, don't try and fix it unless you are competent at electronics and soldering—just send it back to the manufacturer or dealer.

Dry joints are another common problem, especially in touring equipment. These are very common in touring equipment because they are caused by general wear and tear, loading electronics into flight cases and putting them into the back of a truck for them to go bouncing down the road. You can identify a dry joint by an intermittent fault; when looking at the PCB itself, you might be able to see that a solder joint has physically come loose. Fortunately, this problem is easily fixed with a soldering iron. However, if the circuit has been completely detached from the board, this is a more complicated problem; in that case, you are probably better off simply sending the piece of kit back.

CURRENT AND AMPERES

A *current* is the flow of electrons through a circuit. Just like a current in a river, an electronic current can be very strong or fairly weak. We measure this current in *amperes*, or, as they are more commonly called, *amps*. (Don't confuse this term with the equipment called amplifiers; amps measure the rate of current flow in electrical circuits.) Amperage can be measured in amperes or milliamps, which are one-thousandth of an amp.

AC/DC

In addition to being one of the most famous rock bands of the 1990s, AC/DC also refers to types of current. There are two types of current: alternating current (AC) and direct current (DC).

Alternating current changes direction 50 or 60 times a second, meaning it flows backwards and forwards; this is the kind of power you get from the electrical sockets in your house. If you look at an AC voltage on an oscilloscope, you will see that it looks exactly like a sine wave in audio (or a distorted sine wave). That's because it is—and because it is a sine wave, it oscillates in the same way.

Direct current, on the other hand, is the steady flow of electrons in one direction, and it is what *all* electrical devices use. The power supply inside an electrical device converts the AC power supplied from the wall outlet into DC power that the unit can use. A battery is also a DC power supply.

VOLTS

ElectroMotive Force (EMF) is the potential force with which electrons are pushed through a circuit. We measure this potential in *volts*. The more volts supplied by a circuit's power supply, the more potential power you will be able to get out of your circuit.

In some cases, you may need to use your multimeter to measure volts. On the meter, you will see two acronyms: VDC and VRMS. Because electronic circuits work on a DC power supply, you need to use volts direct current (VDC); for other power sources, you need to use volts root mean squared (VRMS) or volts AC (VAC). We use VRMS to measure the root-mean-square level of an AC voltage, which is approximately its average value over time.

WATTS

A *watt* is the unit by which we measure electrical power produced or consumed. Watts, amperes, and volts are intrinsically linked; a watt is defined as:

$$1 \text{ volt} \times 1 \text{ amp} = 1 \text{ watt}$$

You should also be familiar with milliwatts, which are one-thousandth of a watt, and kilowatts, which are 1,000 watts.

OHMS Ω

The ohm is a measurement unit of resistance or impedance. In other words, it measures how easily electrons flow through the circuit. A lower number of ohms means a lower resistance.

Ohms are one of the most important measurements in electronics. Every piece of equipment we use has some input or output resistance or impedance, ranging from speakers and microphones to amplifiers and headphones.

Going back to our running track analogy, you can think of it this way: Every time you add a hurdle, it slows the speed of a runner. Anything, whether it is a light bulb, a fader on your mixer, or even yourself put into an electrical circuit has some kind of resistance.

Some engineers manage to go through their entire career not really knowing what an ohm is or how to work with ohms in their system. You can get by with this lack of knowledge if you just want to mix—but when there is something wrong with the PA system and no one knows what's going on, you'll have a problem. Thus, this subject is really important for all types of audio engineers, whether you are in a studio or workshop or are working live. Understanding ohms will help you wire speakers together correctly. It could also help you understand why amps blow up and smoke out the dressing room. You can use this knowledge every time you go to work.

To understand what an ohm is, let's take a look at Ohm's Law.

Ohm's Law

Ohm's Law states the following: voltage = current × resistance, or resistance = voltage/current. Resistance (in ohms) is voltage (in volts) divided by current (in amperes). In equation form, Ohm's Law is $V = IR$, where I = current in amperes.

To understand this law more thoroughly, let's look at some practical examples that you may encounter in your day-to-day audio experience.

Let's assume that you want to push the master fader up on your mixer by 6 dB. By doing this, you are doubling the voltage, and by doubling the voltage, double the current has been pushed down the line toward the amplifier. Now, let's assume our loudspeaker is 8 ohms. Thus, at 12 volts, our current would be 1.5 amps: 12/8 = 1.5.

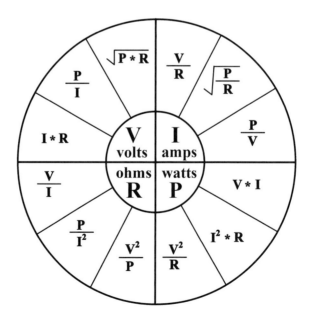

FIGURE 3.1
A rather handy chart to have when working with Ohm's Law. This shows how we can rearrange the letters to find the answer when we have other values. The letters in the middle of the chart represent the end result. The *P* stands for power, so this is your watts. So when we want to find out what the resistance is we can us the pie chart to see that *R* = *V*/*I*.

Thus, using Ohm's Law, you can determine that the power going to our speaker would be:

$$12 \text{ volts} \times 1.5 \text{ amps} = 18 \text{ watts}$$

When you double the power (watts), the signal level goes up only 3 dB, not 6 dB. When you double the voltage, the signal level goes up 6 dB.

It sometimes happens that one side of a system is quieter than the other. Remember: Everything in an electrical circuit has an impedance of some sort, even cable; thus, on a long cable run from an amplifier to a speaker, you could lose as much as 2 ohms. If you're running from a 100-watt, 4-ohm amp into a 4-ohm speaker, the impedance on the speaker and cable would be 6 ohms, giving you only 50 watts of power.

Here is another example: A 2-ohm cable passing 2 amperes of current loses 8 watts of power. Thick (low-gauge) speaker cables have less resistance than thin cables of the same length. So use thick cables (at least 12 gauge) to avoid losing power in the cables.

The higher the resistance of a speaker cable, the more power you lose through that cable, by the equation

$$\text{Power} = \text{current} - \text{squared} \times \text{resistance } (P = I^2 \times R)$$

With this in mind, you should understand that it's always important to make sure you have the same length and thickness of cable on both sides of a system.

On short runs, different-length cables might not be very noticeable—but when setting up larger PA systems, it will have an effect.

Resistance and Impedance

When we place resistors or wires into a circuit, we are adding resistance to the circuit. Even the cable that is connecting the two poles of your power source to the circuit has a resistance, and all real-life circuits have some resistance. (Some circuits created in laboratories use superconductive material that offers practically no resistance, but this is irrelevant to real-life sound work.)

Although the terms *resistance* and *impedance* are often used interchangeably, there is a slight difference between the two. *Resistance* is opposition to direct current (DC), whereas impedance is the opposition to alternating current (AC).

Impedance varies with frequency; resistance does not. Loudspeaker manufacturers specify the average impedance of a loudspeaker over a wide frequency band.

Load

Now that you understand what ohms are and have seen how they can be used in an everyday live audio environment, let's explain the idea of *load* on a circuit. *Load*, put simply, is the impedance in ohms that a circuit has to drive, such as a loudspeaker impedance. To understand this a little more clearly, think about a heart pumping oxygen through a body. In a resting state, the need for oxygen is lower than if the body were working. When you start to move your body, however, the various muscles require more oxygen, so the heart pumps faster to get the oxygen to the places that need it. In other words, the load on your heart increases.

When we were talking about speaker cables having an impedance of two ohms, this is what we call *cable load*. This applies to all types of cable, even if they are oxygen free.

Never use a speaker with a lower impedance than a power amplifier can handle. For example, never put a 2-ohm speaker on an amplifier with a minimum impedance rating of 4 ohms. If you do, the amplifier probably will overheat and may fail and maybe blow up.

SERIES CIRCUITS

A series speaker circuit is a circuit in which speakers are connected from the + terminal of one speaker to the − terminal of the next speaker. The current flows through each speaker one at a time, but the electrons are moving so fast that you will not hear any delay. The more speakers you have in series, the higher the impedance of the total speaker load. For example, if you have two 8-ohm speakers, the load on the amp will be 16 ohms.

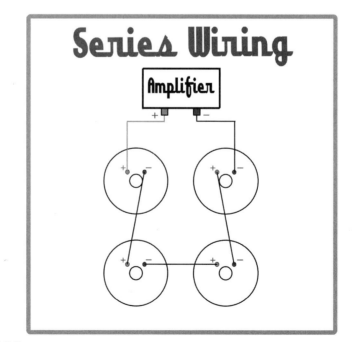

FIGURE 3.2
This is how a series speaker circuit would look.

PARALLEL CIRCUITS

A parallel speaker circuit is a circuit in which the + terminals of the speakers are connected together, and the −terminals of the speakers are connected together. The effect of this is the opposite of a series circuit: The impedance decreases. Thus, two 8-ohm speakers wired in parallel create a 4-ohm load on the amp.

It's important to wire the speakers to obtain a total load that matches the amplifier impedance spec. This makes the amplifier work most efficiently and thus produces the maximum output possible. For example, if you're running an amplifier designed for a 4-ohm speaker load and your speakers are 8-ohms impedance, the amplifer might overheat.

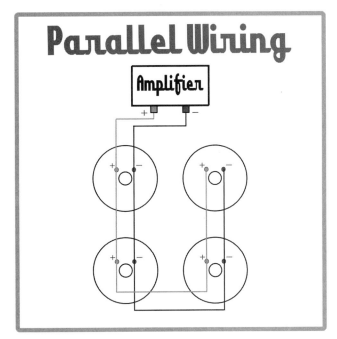

FIGURE 3.3
This is how a parallel speaker circuit would look.

Power Capacity

The power *capacity* of a circuit is the maximum power it can handle without burning out. You should always know the maximum capacity for any given circuit; fortunately, it's quite easy to figure this out. To begin, remember that you should always give yourself a safety zone to work within; the power used in a circuit should be less than 80% of the maximum power it can handle. Referring back to Ohm's pie chart in Figure 3.1, you can see that, in order to determine how many watts a circuit can have, all you have to do is multiply the volts applied to the circuit by the amps that run through the circuit from its power supply.

A normal household socket in the UK has a voltage of 230 volts and an amperage of 13 amps, which results in a maximum wattage of 2,990 watts. This is the maximum power that can be drawn by the power circuit. However, when working with circuits, you should work within a safety zone (just to make sure nothing melts). In this case, then, you should consider the maximum power draw to be about 2,300 watts (80% of 2,990).

Remember: The impedance of a circuit or speaker is the load that it places on the voltage source or power source.

CHAPTER 4
Power and Electricity

Power can be one of the most difficult parts to understand when setting up a sound system. Understanding just the very basics of how power is supplied and distributed around the system is very important and can help find a quick solution to any sort of problem. At this point, we have already discussed the basics of electronics, which makes the jump to power a fairly simple one. The parameters are exactly the same: Electrons move, causing things to light up and make noises.

We have discussed most of the basic physics in previous chapters, but we should explain a couple more fundamental issues before we get to anything else.

FREQUENCY

Recall that when we looked at alternating current from a power outlet, we saw that it looks exactly the same as a sine wave in audio. And, as in audio, the sine wave has a frequency that is set by the power supplied to the sockets. Worldwide, most domestic sockets run between 50 and 60 Hz. You will find that most pieces of equipment can handle this frequency range.

GROUNDING

Grounding, or earthing as it is known in the United Kingdom, is primarily a safety feature of an electrical circuit. Electricity is attracted to the earth, and all electrical current tries to find the easiest path to get there. To make the grounding system work, we drive metal rods into the earth, and the electrical circuit is connected to it. This applies to any circuit; if it is in your home or venue, your electricity suppler will have a ground connected somewhere nearby. If you are using a generator outside, then you'll probably see a bit of cable attached to a metal pole just next to it. If a fault occurs in an electrical circuit, this grounding system will protect people, animals, and plants from getting a potentially deadly shock of electricity, or creating an electrical fire that could burn your house down. A ground rod is used to dissipate electrical charges from faulty currents, static buildup, and lighting strikes. It may also reduce electromagnetic and radio frequency interference.

Live Audio.

All electrical systems need a reference voltage; a good reference voltage is 0 (zero) volts. Ground is defined as the 0-volt reference in a circuit. In AC power systems, we use the earth as our reference voltage because it almost always remains at 0 volts. The voltage at each point in an AC electrical system is measured relative to the earth ground. In an audio circuit, the voltage at each point in the circuit is measured relative to a 0-volt reference called ground, whether or not it is attached to the earth.

Buzzing, Humming, and Ground Loops

Occasionally, you might plug in your system and turn it on, only to hear a buzz or hum at 50 or 60 Hz or their harmonics. This unwanted sound is usually caused by a ground loop. (As you will recall from the earlier discussion, AC power has a frequency of either 50 or 60 Hz.) You'll never know when or where a ground loop will happen, though they are more common in older venues, where the power systems are aging.

The physical cause for a ground loop is generally two or more devices connected to the ground by two different paths. For example, the PA system is connected to one power ground (via the ground pin on the AC plug), and the backline is connected to a different power ground at a different voltage (via the ground pin on its AC plugs). When you connect a bass guitar and its amp to the PA system using a DI (Direct Injection, see page 179) box, both the bass amp and PA are grounded, but by different outlets that may be at different ground voltages. You must isolate one of the grounds to eliminate the buzz. The majority of buzz or hum in a PA system is caused by bad grounding.

Another cause of hum is the pickup of radio waves or electromagnetic interference by an audio cable. Long cables can pick up all sorts of unwanted noise from objects that are emitting EMF or RF (lights with dimmers or radios, for example). This is because the cable acts like one long antenna that picks up noise and puts it into your audio path. One way to combat this effect is to shield all cables (except speaker cables). A *shield* is usually a mesh of wire that is wrapped round the signal wires inside a cable. If you strip the cable back, you will find one or more signal cables—usually color-coded—and a mesh of wire wrapped around them; this is the shield. The shield is connected to the audio equipment's chassis ground, so the unwanted noise that is captured by the shield is sent to the ground, keeping your signal cleaner than it would be without it.

DANGER, DANGER, HIGH VOLTAGE . . .

As we've seen previously, the flow of electrons is measured in amps. Each AC outlet (socket) has a maximum amperage rating that it can produce. And each circuit has a maximum amperage that it can handle without overheating. Most domestic sockets throughout the world have a current rating of 10–20 amps.

The lower the amperage, the less the flow of electrons (the less the current). Large professional power supplies may produce currents that can kill you.

Obviously, if you have a problem with a power supply, you should always seek help from someone who is trained in this field.

Your power supply should have some safety systems in place, so the power is cut off when something goes wrong. There is usually a circuit breaker that will turn off the power to a specific area or socket(s). There will also be a main supply switch, which will cut off the power to the whole supply (if necessary). Always make sure that everything is connected to a trippable circuit. If something goes wrong, you want to be absolutely sure that the power will be cut. Check out this picture of the power supply we had to use in Brasilia when I was down there with Anathema. This wasn't a particularly healthy looking situation, so my advice is to let a professional electrician deal with it.

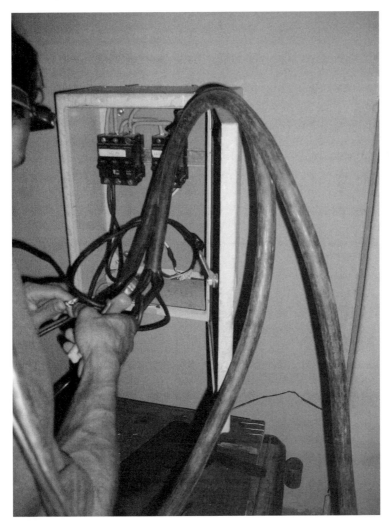

FIGURE 4.1
Fuse box from the Anathema tour when we hit Brasilia.

TRANSFORMERS

A transformer is an electric device consisting of two or more wires wound on the same iron core. The two windings are called the primary and secondary winding. If the secondary winding has more turns than the primary, an AC voltage applied to the primary will be increased at the secondary. This is called a step-up transformer. If the secondary winding has fewer turns than the primary, an AC voltage applied at the primary will be less at the secondary. This is called a step-down transformer.

One purpose of a transformer is to make audio equipment compatible with different mains voltages from country to country. When you are in North America, you will have a 110-V main power supply; thus, if you're coming from the UK with equipment that only works on the UK main voltage, which is 230 V, it won't work. However, if you get a step-up transformer that steps up to 230 V, and plug the 110-V supply into it, you will get 230 V out the other end, and the equipment will work. Many pieces of audio gear have a power transformer with several windings that handle a wide range of AC mains voltages.

Remember always to read the labels on your kit before plugging anything in. Specifically, you should check the input voltage for the equipment. However, a lot of modern equipment accepts input voltages across a wide range now, so you may not need an external transformer. Above all, never plug any piece of equipment into a voltage it does not accept; you could damage a lot more than the equipment you plugged in.

World Power Systems

If you are traveling around the world and taking your own equipment with you, you must check that you will have the voltage your equipment needs. These days, most power supplies run between 110 V and 240 V, which covers most of the world. Some, like Japan, even go down to 100 V. When checking your equipment, you need to look at the AC voltage required, which is usually found next to the AC power socket on the device, or in the information box on the underside or back.

SOLVING PROBLEMS RELATED TO AC POWER

A major problem that can be potentially fatal is getting shocks from microphones. Shocks happen when there is a difference in voltage between two connected objects; it usually happens when you have a guitarist who is using a mic. In this case, the mic is connected to the mixing console, and the guitar is connected to the guitar amp. Those two devices may be at different ground voltages, causing a shock when the guitarist touches both the guitar strings and the microphone.

When trying to troubleshoot this type of problem, you must consider all possibilities. First, you should unplug the amp and make sure it is connected to the same circuit as the PA system. All of your backlines need to be connected to the same power circuit as the PA because they are connected to the PA system. Older amps will sometimes have a polarity switch, which swaps the hot and neutral AC wires. Sometimes you might find that the power socket is wired differently. For example, in some older amps, where the chassis is connected to the neutral (because the neutral is another way for the flow of electricity to get to the ground), the live wire could be connected to the chassis, which would be bad because you would be connecting the mains to the outside of the amp and if you touch it, you're going to be toast.

Another way of trying to fix shocks from a microphone—and this a more of a last resort—is to get the electrical current to go to another point in the power circuit. This is called *shunting*, and you can do it by using a DI box. Run the guitar through the DI box, into the amp, and then connect the DI box into the stage box or desk. You aren't trying to isolate your guitar from the amp, but you are trying to connect all the grounds together. Make sure that the earth lift on the DI is not activated so the earth connects to the PA. Hopefully this will be enough to push the current down to ground.

Along with the ground hum we spoke about earlier, these are just a few examples of problems you might encounter during your time as an audio engineer, and some simple solutions that could fix them. As a general rule, the best thing to do is to carry a multimeter in your toolkit at all times. You can also get hold of plug testers that plug directly into the power source, which will test if the earth, live, and neutral are connected properly. Always remember, though, that you should have a professional electrician sort out any problems that don't have a simple and straightforward solution.

CHAPTER 5
Advancing the Show

Before a tour has even started, part of your job as an enthusiastic audio engineer is to *advance the shows*, which means that you create documents of your band's technical specifications (or "tech specs") and then send them over to your tour or production managers so they can send them to the venues and/or promoters, along with any other documents they need. You need to make sure you have received the tech specs from each of the venues or the PA companies that are supplying the gear. This process needs to start sometime before the first show.

As part of advancing the show, it is also your responsibility to go through the venue specs and make sure everything you will have at the venue will work with your band's setup. For example, are there enough channels in the multicore (snake) for you to send all your mic signals to the front of house? Do you have enough tie lines (which are a way of sending a signal from FOH) to send everything you need to stage? Have you used the console before, and are you familiar with this kind of PA system, if you aren't touring your own? And even if you *are* touring your own PA system and consoles, you need to liaise with the PA company to make sure it has everything from you that it needs to put the whole system together.

BUDGET

Although you won't be expected to prepare a budget for your kit, and you probably won't be involved in that process at all, everything comes down to budgets at the end of the day. If you want something, someone has to pay for it. On smaller tours there is usually not even enough cash to get you around the country, let alone pay for the latest gear. So, when there is something you need, speak to the people who write the checks, and get them to sort out who is paying for it.

Remember: a big part of your job is to get the best out of what you are given.

TOUR AND PRODUCTION MANAGERS

Most of the time, you'll have a tour or production manager who will be dealing directly with venues and promoters. Create your channel list, stage plan, equipment rider, and so on, and send them over to the tour or production manager to send off to the promoters with the rest of the rider. All of this content is part of the contract for the show, so it's really important that everything gets sent over in full and together.

Tour and production managers are your direct contact, and you will work closely with them. Everything you do should be run through them, unless you've been told otherwise. It is also their responsibility to get you the tech specs for the venues you are going to be working in. If you don't have them, let your *tour manager* (TM) or *production manager* (PM) know so that they can get them for you. They are very busy people, and a tech spec from a gig in Birmingham isn't going to be at the top of their priority list—so sometimes a gentle reminder is necessary.

CHANNEL LISTS AND STAGE PLANS

As part of any good advance, you are going to need to put together what we call a *channel list*. This is the list of channels that you require on the mixing console to be able to get everything on stage through the PA system. The other thing that needs to be put together is a *stage plan*. This is a top-down view of all the positions of the band and their instruments on the stage.

Channel Lists

I added two examples of channel list I've made in the past (Figures 6.1 and 6.2). One of the key points when piecing your channel list together is to remember that you know what is happening on stage; the venues won't, so adding as much information as you can, in as simplified a form as possible, will help everyone involved. Always include your name, position, phone number, and e-mail address—there are always questions that need to be answered. You should also include a version number and/or a date on the document. You may also want to take advantage of systems that allow you to share documents, enabling multiple people to edit them without creating multiple versions (for example, Google Docs). This way you'll be sending out the most up-to-date version of the document every time. I can't count the number of times we've turned up at a show and the venue has a channel list that is 9 months old because the agent didn't send over the latest version. This procedure should hopefully eliminate this problem.

As you can see from the figures, these documents are split up into columns listing channel number, instrument, type of microphone, any inserts required, whether phantom power is required, what type of stand you need, and any other notes that might be relevant to your setup. On the La Roux channel list, you can see that I have color-coded a section called *La Roux loom*. Now that we carry our own mics cables and looms (we also use the term *lines* to mean the line from the mic to the console) for the stage, we have a patch bay on the back of the equipment that runs straight to our own stage box for the mic lines. This is really just for our benefit because of the ways the lines have had to be labeled. But having your stage positions labeled on your channel list will help

LA ROUX

Updated:
1st July 2010

Channel List

Ch	Instrument		LR Notes	Input	La Roux Loom	Insert FOH	+48v	Stand
1	Kick		HD 1	Own DI	Key Rack - 1	Comp		
2	Snare		KIT	DI	DRUMS - 3	Comp	Y	
3	Sn + Hts		HD 2	Own DI	Key Rack - 2	Comp		
4	KIT	L	KIT	Own DI	DRUMS - 1	Comp	Y	
5	KIT	R	KIT	Own DI	DRUMS - 2	Comp	Y	
6	Perc	L	HD 5	Own DI	Key Rack - 5			
7	Perc	R	HD 6	Own DI	Key Rack - 6			
8	Elly Perc		KIT	Own DI	DRUMS - 4			
9	Bass		HD 3	Own DI	Key Rack - 3	Comp		
10	Moog			Own DI	Key Rack - 11	Comp		
11	Bass		Lap 1	Own DI	Key Rack - 12	Comp		
12	Bass Guitar			DI	Key Rack - 19	Comp		
13	Acoustic Guitar			DI	Key Rack - 21	Comp		
14	Melody		HD 4	Own DI	Key Rack - 4	Comp		
15	Track	L	HD 7	Own DI	Key Rack - 7	Comp		
16	Track	R	HD 8	Own DI	Key Rack - 8	Comp		
17	Synth	U/S	LAP 2	Own DI	Key Rack - 13			
18	Nord	U/S		Own DI	Key Rack - 14			
19	Synth	S/L	LAP 3	Own DI	Key Rack - 15			
20	Nord	S/L		Own DI	Key Rack - 16			
21	Elly Vox	Centre		Radio Mic	Straight to Split	Comp		Own Stand
22	BV		HD 9	Own DI	Key Rack - 9			
23	BV		HD 10	Own DI	Key Rack - 10			
24	Mickey Vox	U/S		e935	Key Rack - 20			Own Stand
25	Spare Radio Vox	Centre		e935	Straight to Split			
26	Click			XLR	Key Rack - 17			
27	Click 2			DI	Key Rack - 18		Y	
28	AMBIENT			Cond	DRUMS - 12			Short Boom
29	AMBIENT			Cond	Key Rack - 24			Short Boom

We will be carrying all our own Mics, DI's and Stands where stated.
We are carrying our own In Ear Monitor Systems for each member of the band.
We will be carrying our own monitor console. Please provide XLR tails to patch into this.

FIGURE 5.1
The La Roux Channel List.

the in-house techs run the satellite boxes or looms of mic cables to the right place on stage. (This is especially useful when doing festivals with quick changeovers, so that everybody knows exactly what is supposed to be going on.) Remember: Communication is key to having the whole day running easily and well.

Updated:
30th June 2007

Channel List

Ch	Instrument	Mic		Insert FOH	FOH VCA	Insert Mons	Stand	Notes
1	Kick (Con)	AE2500DE	(Own)	Gate	1 + 7	Gate	Short Boom	
2	Kick (Dyn)	"		Gate	1 + 7	Gate		
3	Snare Top	ATM650	(Own)		1 + 7		Short Boom	
4	Snare Bot	AE3000	(Own)	Gate	1 + 7	Gate	Short Boom	
5	Snare 2	ATM350	(Own)				Clip (Own)	
6	Hi Hat	AT4041	(Own)		7		Short Boom	
7	Rack Tom	ATM350	(Own)	Gate	2 + 7	Gate	Clip (Own)	
8	Floor Tom	AE3000	(Own)	Gate	2 + 7	Gate	Clip (Own)	
9	Floor Tom	AE3000	(Own)	Gate	2 + 7	Gate	Clip (Own)	
10	OH (SR)	AT3060	(Own)		7		Tall Boom	
11	OH (SL)	AT3060	(Own)		7		Tall Boom	
12	Bass Di	Active DI		Comp	3 + 7	Comp		
13	Bass Mic	ATM250	(Own)		3 + 7	Comp	Short Boom	
14	Guitar	AT4050	(Own)		4 + 7		Short Boom	
15	Guitar	AT4050	(Own)		4 + 7		Short Boom	
16	Keys L	Active DI		Comp	5 + 7			
17	Keys R	Active DI		Comp	5 + 7			
18	Whurlitzer	Active DI		Comp	5 + 7			
19	Baritone Sax	Pro 25AX	(Own)	Comp	6 + 7		Short Boom	
20	Tenor Sax	Pro 25AX	(Own)	Comp	6 + 7		Short Boom	
21	Trumpet	Pro 25AX	(Own)	Comp	6 + 7		Tall Straight	
22	Flute	AE5100	(Own)		7		Tall Boom	
23	Key Vox	ATM610	(Own)	Comp	8		Tall Boom	
24	BV1	ATM610	(Own)	Comp	8		Tall Boom	
25	BV2	ATM610	(Own)	Comp	8		Tall Boom	
26	Amy Vox	ATM710	(Own)	Avalon 737 + BSS DPR901	8	Comp	Tall Straight with Round Base	Long Cable
27	Spare	ATM710	(Own)	Comp	8	Comp		Long Cable

FIGURE 5.2
The Amy Winehouse Channel List.

Stage Plan

The key to a well-put-together stage plan is to keep irrelevant information out and retain just the basic information needed for the stage. On the two examples of stage plans above, you can see that everything is clearly labeled; you can see where the power drops are, what AC mains voltage is required (this

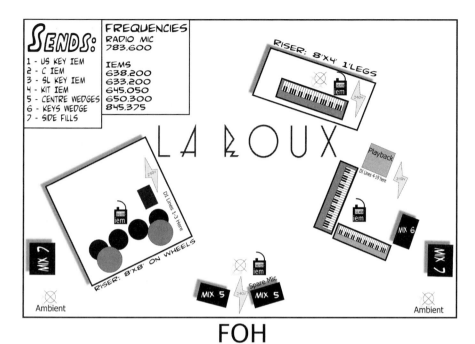

FIGURE 5.3
The La Roux Stage Plan.

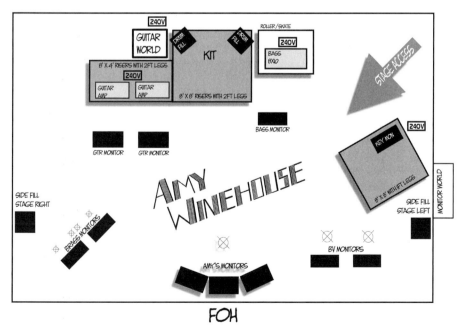

FIGURE 5.4
The Amy Winehouse Stage Plan.

information is important if you are traveling internationally), what risers are required and how tall they need to be, and whether wheels are needed.

The circle with the X through it is the technical drawing symbol for a microphone, so I have used this symbol to show the location of the vocal mics on stage. You don't need to add the rest of the microphones on here because they are written on your channel list; as long as you have the placement for your kit, the engineers should know where to run the lines and how many you need.

PA SPECIFICATIONS

Whether you are working for a band, venue, or PA company you should be able to put together a *PA Spec*. A PA Spec is a document that holds all the key information about what type of equipment you would like to use and/or bring into the venue (if you are traveling with the band), or what type of equipment is in the venue (if you are working in house).

From a touring artist's point of view, you should be specifying what equipment you want to use or are bringing into the venue like desks and PA system, and how much and what type of outboard equipment, such as gates, compressors, delays, and reverbs you need to run the show. Putting this information together will help the venues and promoters get the right equipment for every night, and if there is going to be a problem getting hold of certain elements, they can find out if you have any alternatives.

The way the PA spec should be split up is the same no matter if you are the band's audio engineer or the in house PA tech. It's always a pretty good idea to start with the loudspeaker section: List what type of box you have, how many there are, and what they are used for.

Here is the PA spec I used for Amy Winehouse:

FOH

System: D&B Q1 / J8, L.Acoustics V-Dosc / Funktion One Res5
    The System must be capable of producing up to 120 dB SPL in a frequency range of 30 Hz–20 kHz and must be able to produce equal sound dispersion throughout the auditorium. The system must be electronically crossed over.

Unacceptable PA Systems: Peavey, Old Martin, and Home Made Boxes

Desk: MIDAS H3000
    MIDAS H2000
    MIDAS XL4

A fully working Parametric EQ is required for each channel, with Q, pads, Hpf, Phase Reversal, Phantom Power, and 8 Auxiliaries.

Must NOT be on a riser, on, or under a balcony.

It must have at least 48 mono channels and 6 stereo channels with fully functional parametric eqs on all channels, 8 VCAs, 8 subgroups, and 8 mute groups.

Unacceptable Desks: Yamaha, Allen & Heath, TAC, Crest, Mackie, Hill Audio

Control: Lake Contours on all outputs to the speaker system (will require separate
control for flown, ground, delays and infills).
Stereo Graphic EQ KT DN370 inserted on masters L & R, Sub included
Klark Technic DN6000 with calibrated Microphone
8 Channels of Compression: DBX 160A
6 Channels of Gates: 3 × Drawmer DS501
2 × TC Electronic M2000

1 × Roland SDE 3000
1 × CD Player
Coms To Monitors
Talk to Stage (Mic with Switch)

NO BEHRINGER or SAMSON ANYWHERE IN THE SYSTEM!!

As you can see from this spec, it's fairly basic. It states what I would like, alternatives if they don't have what I am after, and what I'm really not willing to use.

If you are putting together PA specs for a venue, then you need to list all the equipment you have in the venue so that this can be sent off to all the bands that are going to be coming through the venue. Take a look at the following layout; you'll see how I've split each part of the whole PA system up into sections. This will help whoever is looking through the spec to quickly see what you have and how much of it there is. If anyone is looking for a piece of equipment, then it's easy to find.

FOH Loudspeakers

D&B Audio Q1 Line Array
5 × Q1 Mid/Hi – Main PA
1 × Q7 Infill
6 × Qsub

FOH Speaker Management System

BSS FDS 366 Omnidrive Compact Plus

FOH Amplification

5 × Camco Vortex 6 Amplifiers
2 × Camco Vortex 4 Amplifiers
4 × Camco DX 24 Amplifiers
6 × Camco DX 12 Amplifiers

FOH Console

1 × Midas XL200 Console, 44 mono, 6 Stereo

FOH Processing

1 × BSS FCS-966 Dual 31 Band Graphic EQ
2 × BSS DPR502 Quad Gate

2 × BSS DPR402 Dual Comp/De-esser
2 × XTA C2 Stereo Compressor
1 × XTA G2 Stereo Gate

FOH Effects

1 × TC Electronic D2 Multitap Digital Delay
1 × Yamaha SPX1000 Multi Fx
1 × Yamaha SPX 990 Multi Fx

Monitor System

8 × Martin LE400C Floor Monitors In Matched Pairs
2 × Martin H3 Sidefills
2 × Thunder Ridge Tri-Amp Drumfills

Monitor Speaker Management

1 × Yamaha DME 32 Configured for 8 Way Bi-Amp Foldback System

Monitor Amplification

4 × Yamaha PC4800 2 X 800 W Power Amp
4 × Yamaha PC9500 2 X 1400 W Power Amp

Monitor Console

1 × Soundcraft MH3 40ch Console

Monitor Processing

4 × BSS Fcs-966 Dual 31 Band Graphic EQ
2 × Drawmer Quad Gate
2 × BSS DPR402 Dual Comp/De-esser
2 × Behringer MDX440P Multicom Pro
1 × Behringer XR440P Multigate Pro

Monitor Effects

1 × Yamaha SPX1000 Multi Fx
1 × Yamaha SPX 990 Multi Fx

Line System

1 × 40+10 Multicore
1 × 16 Way Satellite Box
1 × 12 Way Satellite Box
1 × 12 Way Tie Line (FOH – Stage)

Microphones

1 × SHURE BETA 91
2 × SHURE BETA52
6 × SHURE SM57

6 × SHURE SM58
4 × SHURE BETA 57a
7 × SENNHEISER E604
2 × SENNHEISER E606
1 × AKG D112
3 × AKG C418PP Clip on drum condenser
4 × AKG CK391 Condenser
1 × BEYERDYNAMIC M88
8 × BSS AR133 Active DI

Microphone Stands

11 × Beyerdynamic Tall Boom Stands
11 × Beyerdynamic Short Boom Stands

After you have written your spec, you should include any other additional information you think might be important, such as how the PA system might be split up into zones and what type of DJ equipment you have.

Venue Specifications

So that takes care of the PA systems, but sometimes , if you are an in-house tech, it could fall on your shoulders to put the other specs together for the venue. Your PA spec can be expanded to include information about lights, catering, merchandise, and other details.

When you're putting together venue specs, there are some very key things you should include, so that artists know what to expect well in advance of a show. For example, you should include all the equipment you have, information about any pieces of equipment that are going in for maintenance, and details about whether there are any changes to anything else that matters (such as load-in or the capacity of the venue).

When putting the layout together, make sure that you have all the important information about your venue right at the top of the first page. These would be things like the name of the venue, address, and contact details. Also make sure that you state the address of the load in, in case it is different from the main venue address (for example, the street behind the venue or to the side of the venue), along with contact details of the person in charge of loading the gear in, and where to park the bus/van/truck. You should state whether they will have to tip and go, or if there will be adequate parking for the duration of the show. Another handy thing to do is use a footer on each page of the document with page number and venue name on it. That way anyone printing anything out knows if they have lost a page, or where it goes if they find the page.

After you've got that all laid out, split your spec into different sections. Contact details should be right at the top underneath the important load-in information. Include all production staff members that are necessary to the production—for example, promoter, production manager, head of sound, and head of lights in that section.

Next, provide all the venue details. Start by listing the capacity of the whole venue; then, if applicable, split it up into the capacity of the different sections of the venue (floor, balcony, bar, etc.). List any curfews that might be in place. For example, you may have a sound curfew until 5 p.m., or you may only have a venue license until 10.30 p.m. This type of information must be included and is vital for the touring personnel to know when they are planning load-ins and load-outs.

Next, look at the venue's dimensions. Adding the internal schematic drawings of the building is always a good idea, for this will help incoming production to fit into your space, but if you don't have them, you should at least include the dimensions of the room. Make sure to clearly indicate where the FOH mix position is (if you don't have drawings showing the mix positions, make sure you state whether it is audience left or stage left), and whether the mix position is fixed or can be moved. State the distance between the mix position and the stage.

Details for the stage should include its height, width, and depth, stage and floor clearance (the distance from the floor or the stage to the ceiling), and details on the wings of the stage (how big they are, whether anything is stored there, and information about the locations of permanent installations such as dimmers and monitor consoles). In addition, information about points is absolutely essential; *points* are where you hang PA, trusses, cables, and more—basically, anything that goes in the air is hung from points. You should state where they are, how many there are, and how much load they can take (adding a schematic drawing is best here). If you don't have points in your building, but gear still needs to go in the air, you'll probably need ground support. This should be handed over to the promoter and the production manager to sort out. After this, add any information you have about a crowd barrier at the front of the stage—that is, whether you have one and, if so, its distance from the stage.

Once that section is complete, you should address the practicalities of getting the artist's equipment in and out of the building. All the additional information about loading into and out of the venue should be written here. You should include how long the load-in is, whether there are stairs or an elevator, where the storage area is, and how much storage area there is (for empty cases, other equipment that might not be used during the show, etc.).

Power Specifications

I placed power specifications in an additional section because it's quite important to get this material right; otherwise you might end up in a cloud of smoke and with no equipment working. There are two sides to this: the power that is required from the venue as a touring production and what power the venue has.

Let's take a look at the touring side first.

Make sure you have specified exactly what power you need, including the voltage, amperage, and positions of the power sockets. It's a good idea to add this information to your stage plan, showing the exact position of the power drops. If you are writing a spec for a PA system, make sure that the venue has the correct power to suit your needs and that you have enough adaptors and cabling to get the power where you need it.

This is the information we used to give on the Winehouse tour:

Power

P.A. 63A 3 Phase Stage Left, Camlocks or Cee Form.
Lighting 100A 3 phase Stage Right, 125A Camlocks or Cee Form.

Now let's look at the information you should have about power in the venue. State what circuits you have ready—for example, lighting, PA, on-stage, and any others. State what amperage each circuit is, how many phases each circuit has, and what type of connector it uses.

Flexibility

Everyone has budgets to work within and it isn't always possible to get the right kit. You should be flexible, and give two or three alternative choices of equipment. If something is so important that you absolutely must have it, you should probably bring it along yourself.

PICK UP THE PHONE

Even if you have read through all the specs and worked out that the production will be fine, you should still pick up the phone and speak to someone. It's all too easy for the spec to be sent but with no follow-up communication—and suddenly the band turns up to the venue to find that gear is missing because the band thought the venue was getting it and the venue thought the band was getting it.

CHAPTER 6
Rehearsals

Rehearsals are an incredibly important time for a sound engineer. They are when you might see the gear for the first time, when you first meet the band and the crew, and when you get to learn the music and the songs. With this in mind, this chapter is about everything that happens around the time rehearsals are taking place.

THE MUSIC AND THE ARTIST

Your band will probably have been in rehearsals for some time before you even get involved. The first thing you need to do before you even walk in through the door of the rehearsal studio is to get hold of the artists' music and understand it. This may sound fairly obvious, but what isn't as obvious is that you also need to understand the mental state of the artist when he or she wrote the album (assuming they wrote the album). You need to try to understand the way the music flows within the album, to help you understand how to translate the music into the live environment. The way I approach this is by listening to the album from beginning to end a few times, listening to how the songs flow into one another. Then, assuming the artist has a vague idea of what the set is going to be, I rearrange the songs into the set order. Then I listen to the changes in how the songs work together. You'll be listening mainly for the high and low points of each song, as well as for the relation of each song in the set. Once you have this in your mind's eye, you know the points at which the music needs to be pushed, and needs to be held back. Also when listening to the music, you should take notes. Pay attention to the overall sound of the album or albums. Listen to how the vocal is placed in the mix, and how the bass lines and percussive lines work together. Pick out some key melodies and the obvious parts that could happen within the song and make it interesting. Listen for some more subtle parts to pull out in your mix; they often make a nice addition to the sound, and make your mix stand out. Sometimes it might be best to write these types of notes down, but having everything committed to memory promotes a more organic flow. This all being said, you must also make sure not to overanalyze the music; otherwise, you'll constantly be making changes.

Talk to the artist about what he or she expects from the live sound. Listening to the music beforehand and showing that you have listened will help instill confidence, while also potentially improving their live performance. You may find that the artist wants a very different sound from the album, for example. Sometimes you'll have room to be creative, other times not; but either way, having a good musical understanding of their work will be to your advantage.

The live show is always changing, so you need to note what is happening in the mix. Working out what you are going to do or how you can improve the music is very important at this stage.

SCHEDULE

Making a schedule is essential when rehearsing, especially if you have only one or two days to prepare everything for the tour. You are going to need to work out when you will be able to look at the equipment you are carrying and whether there are going to have to be any last-minute equipment changes. The bottom line is that it's essential to figure out what and when everyone is doing what they need to do. For example, the techs in charge of the backline might have been in for the very first rehearsal to set everything up, but have to come back before you run through any changes since that first rehearsal. It's important to keep this is mind when making your schedules as it ultimately saves time and money—and you won't get in each other's way.

EQUIPMENT: PREPARING THE GEAR

Before you set out on any swashbuckling adventure, you need to make sure all the gear is working correctly and you have all the adaptors you might need. Get an idea of input requirements from the band members so that they can let you know if anything is wrong or if there is anything that could make things a little easier.

SOUNDS AND PROGRAMMING

Using your time effectively at rehearsals can relieve a lot of boredom. You may find that when you are asked to come to rehearsals for the first time the band members are still rehearsing and you have to listen to half-finished songs, and the set might take a whole afternoon to play through. At this point you should be listening to how all the sounds fit together. This is where a solid understanding of different types of music comes in handy. Listen to how all the instruments work together acoustically (that's what you will be micing up, after all), and make sure none of the sounds are clashing. There might be a lot of sounds in the same frequency range; for example, the bass and guitars could cloud up the whole mix. The guitarist wants a nice chunky sound with lots of balls, so you could be looking at frequencies going down to 120 Hz, sometimes lower. In this type of situation, your bass guitar has a big role to play, pushing the energy along with the drums. If the guitar is also going down this low, you'll find a lot of woolly sounds (lower frequencies that cloud your mix) that need to be removed in order to pull some definition back into the mix.

Rehearsals are also a good time to try and find a good stage volume. Remember that the band will probably have monitors on the stage; you need to find a happy medium between what is too loud for the front of house and what is too quiet on stage. You won't truly know what works well until you are in the first venue, but a lot of guitar sounds that are run through valve amps are dependent on how loud the amp itself is. To control the stage volume in the house, it helps to place the guitar amps to the side of the players rather than behind them, and to aim the guitar amps upward at the players' ears. That makes the guitar amps sound louder to the musicians but quieter to the audience.

In addition, if you are fortunate enough to be taking some sound equipment with you, now is probably the time to start making sure you have everything in order, such as programming consoles and effects units. Effects can really enhance your mix, and getting the right reverb or the corrected delay time will help, but make sure you understand the way the equipment works. You will come across delay units that have a tap delay function (a button you can press in time with the music to get the correct delay), that you will be able to use on the fly during the show if needed, but you may also want to ask the band how long they want their delay.

FOOTPRINT

As much as stage sizes change and venues get bigger and smaller, there is nearly always an optimum size that the band prefers. The bottom line is that the band needs to feel comfortable with the size of the stage, and this can be helped along by having the same size *footprint* for every show. In other words, you can help by making sure all of their equipment is the same distance apart from show to show, allowing them to best command the space. Although this may sound like a backline job (and it is, as far as setting up the equipment goes), it's your job as an audio engineer to get the best out of the band on a day-to day-basis. Having a comfortable artist on stage will help you get the best possible mix.

CASES

And now for a bit of housekeeping . . .

Having flight cases for all your equipment is the best way to keep your gear safe and sound and in good working order when traveling. After all, you want to get the best possible sound out of your equipment, so it really should be in the best possible condition. Although this is another backline job, people sometimes need a little push in the right direction to get the ball rolling when it comes to getting cases made. In the early days of a band's career, flight cases may seem like a bit of an overindulgence—but when the amp falls out of the back of the van, it won't seem like that anymore. If you can manage to afford them, cable trucks are also extremely useful; they are a great way to keep all your cables in good working order. They also help you keep track of all the small pieces you have to carry around with you. Over the last year we've been using Samsonite suitcases, which have done the job very well, but now that the La Roux setup is getting bigger and bigger, it is time to retire the good old cases.

When getting cases custom made, think about how you will be lifting them, and make sure you have enough handles and that they are in the right place. Also be sure to keep the size reasonable; you want the cases to fit through an average size door, and oversized cases lead to overfull cases. There will always be times when you have to unload a big trunk at the side of the road in the pouring rain because it won't fit through the average size Swiss doorway, but making good decisions about your cases can make this less painful than it has to be. The weight is also important: 110 pounds (50 kg) is the maximum weight *some* airlines will take when freighting equipment. We were doing some tours in Australia and some of the internal flights would not take any pieces heavier than that. Finally, make sure you decide which cases need wheels. There's nothing worse than having to carry lots of gear when you could be pushing it.

Just remember: "we don't need another hero." Cases can be heavy, and if you hurt yourself lifting something, your tour is over. It's just plain common sense really.

LOOMS AND LABELS

Take the time to make up cable looms (snakes) and label your equipment so that anyone can roll a case in to a venue, put it in the right position, and start plugging things in for you. You will be endlessly grateful for having done this the first time you're late due to a delayed flight or missed train, but it also helps in having a nice and neat stage that will allow a fast and efficient setup. When something goes wrong, you'll be glad you aren't digging through endless knots in cables trying to find out which one you have to pull out. And no matter how hard you try, they always end up in a knot!

If you can, make up looms for any piece of equipment that require multiple mic or ¼ inch cables like the entire drum setup, or keyboard setup—providing you know how the drummer or keyboard player sets up. If the setup on stage is pretty big, it's a good idea to loom as many cables as you can together at the correct lengths, to go from the mic/DI to the satellite box (sub-snake). It'll save loads of time and also keep the stage nice and neat.

A great handy little hint is to number and label all your cases with the name of the band and venue position. This will make sure that the local crew pushing your gear into the gig knows whose cases they are and where each of the cases goes. Make sure you have a list of the case numbers and their contents, which helps make sure that all your equipment stays together. This way, when something gets lost, you will know what it is and you will be able to describe it accurately.

CREATING STAGE INFRASTRUCTURE

Stage infrastructure is put together to make your life easier when setting up and packing down on a day-to-day basis. The infrastructure consists of all the mic cables, power distribution cables, and the stage box(es). Carrying your own infrastructure will ensure that you have everything you need on stage and will require no other equipment or cables from anyone else. You'll be completely self-contained. If you keep all your own equipment self-contained, it will help with faster and more efficient setups. Even though you will have more to do, you won't have to deal with local crew asking general questions, and you will know that everything works. It also helps keep the stage neat and tidy. If you are going to be building this kind of thing with a company that specializes in it, you need to leave at least a month between your initial plans and getting everything delivered to you—sometimes a lot longer, if it's a time of high demand.

There are many reasons to create a stage infrastructure, from making life simpler on stage to avoiding catastrophic equipment failure. To illustrate this point, in a recent experience I had a show where one of the cables to the monitor console was out of phase on the keyboard channels, which are stereo. Because we were running late, soundcheck was a little rushed, so the problem wasn't obvious but caused endless trouble throughout the entire show. We were also prone to a few power issues during this show. The power supply we were attached to was a generator, and it was running a few cycles per second faster than it should have. Remember how we looked at an AC power supply running between 50 and 60 Hz back in the previous chapter on power? Well, this particular power supply was running slightly faster. This wouldn't normally cause problems, but we do have a couple of pieces of equipment that are a little sensitive and refused to work. Take this into account, and think slightly more about protecting sensitive pieces of equipment. Power spikes or complete power failure can be catastrophic for hard disks, so integrating a power distribution system into your stage infrastructure is a good idea. The power should be conditioned so that you are supplying your sensitive pieces of equipment with nice clean power, so there should be far fewer issues to worry about. Another factor in creating these stage systems is that during festival season, it's always nice to use your own system, rather than having the festival tell you what your patch will be. Ultimately, what it comes down to is that, if you can afford it, building a stage infrastructure is a handy thing to do.

Begin by planning the number and location of lines on stage. The setup I am running at the moment with La Roux has 4 lines where the drum kit is, 18 lines coming out of the keyboard rack, 1 backing vocal at the upstage keyboard, a center vocal mic (which is a radio mic, so the receiver is located by the monitor console), and a spare vocal mic; all in all, it's about 24 lines. We also have 2 lines that need to go only to monitors, which are ambient microphones for the band's in-ear monitor systems. For the UK tour, the band decided to add more items: an acoustic guitar, a bass, and a set of steel drums, all meaning another 6 lines. As you can see, the system needs to be diverse to cope with an ever-changing channel list.

To make things fast and easy, we decided to make all the lines from the keyboard rack internal. All we need to do is plug one multipin connector into the back of the rack, and this supplies all the lines from the keyboard rack straight into the splits for monitors and front of house. We decided it would be a good idea to add 10 XLR sockets on the back of the panel, so we can plug external things into the rack as well, eliminating the need to run more mic cable across the stage. This is also where the backing vocal mic on the upstage keyboard will go, and also where we can plug all three extra guitar lines. Over by the drums, we only have 4 lines from the kit itself, but decided to make it a 12-line stage box to accommodate the steel drums over on that side. A spare microphone can also plug into this box.

Next consider the splitter rack. (We'll go into more detail about the mic splitter when we talk about line systems in Chapter 10.) All the mic cables from the stage come into the splitter rack and are then fed out to the front of house console, monitor console, and another split (in case you need to do any recording in the future). On this tour, we decided on a 48-channel split—this is versatile enough that we can expand to a larger setup when needed. Whenever there are more than two splits, they must be made active, which means they are powered and lead to less signal degradation over all the lines. It's also important to make the splitter rack versatile enough so that if the patch changes or a line goes down, you can easily move the entire line or swap the cable. With this in mind, we decided to put in a system of tails that came from the multipin sockets on the front of the rack (coming from our keyboard rack and drum stage box), out of the front of the panels on the side, and then into the front of the splits (using standard 3-pin XLR connectors). Make sure the lid of your flight case is deep enough so that you don't need to unplug it every time you put the lid on.

Consider the power requirements for your equipment. Power must always be very steady, as computers and other delicate machines are negatively affected by variations in power. In other words, you must condition it to a constant voltage and frequency. You should also consider a UPS (uninterruptible power supply), which is similar to a giant battery. It should hold enough charge so that, if the power goes out, you have enough time to save what you need to save and properly shut down all the equipment. If you are using a UPS, you need to know how many watts you're pulling; recall that watts measure how much power is consumed. All UPSs are rated in watts, and knowing your wattage will help you learn how much time you have before the UPS shuts down.

The UPS feeds all the power on stage: the monitor console, the radio mic receiver, in-ear monitor systems, and the console itself. We also needed power for the keyboard rack, the up-stage keyboards, and the drums. We decided that the best plan was to feed the power into the stage box and keyboard rack using a connector called a *powercon*. (These have a higher rated amperage rating than an ordinary plug socket, and lock in position, which makes sure they don't come loose and fall out.) As a precaution, we also added a standard 13-amp UK socket; that way, if any powercons got damaged, we could still use one of our normal cables.

If you're making your own system, you want to make sure it's versatile enough to easily expand without having to get too many things re-done. Getting everything custom-made can be very expensive; most places carry pre-cut strips of rack, which allow you to add as many sockets as you need to. You can also rent these types of systems from various PA companies.

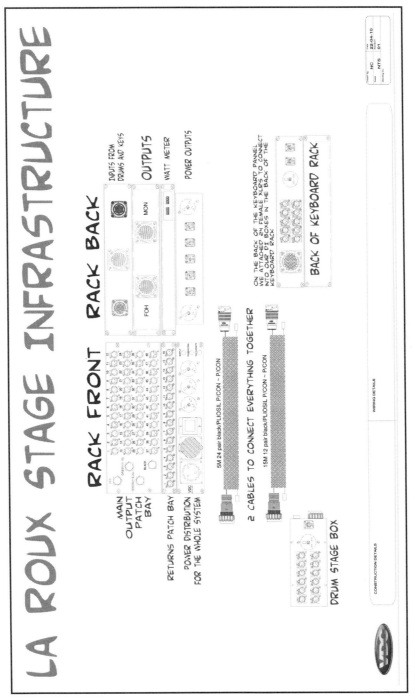

FIGURE 6.1
The La Roux stage infrastructure diagram that Nick Chmara from VDC and I devised.

In Chapter 8 we'll look into calculating the values of amplifiers so you make sure you have sufficient power to run your stage equipment.

FOH CASE

Another important element to think about is what you carry with you in your "front of house case." In this section, we discuss various items that you should always have on hand when on tour. These are the things I carry in my little beige front of house pelican case.

Pencil Case

You might find it useful to use an old mic case to store small items such as pens, white PVC tape, and USB flash drives. (It's now very common to see digital consoles that take USB drives to store data, so it is always best to carry them around with you.)

Sex Changers, Adapters, and Y-Splits

Sex changers are adapters that change the gender of a connector; that is, they change a male connector into a female connector, or vice versa. These can be really handy if you've run tie lines to the stage and need to send another signal; using a sex changer, you can convert the multicore lines so that they send from the desk rather than receive. And it's also handy to have a couple of XLR Y-Splits so that you can split the signal from your vocal mic into two channels of your FOH console; one channel will control the sound going to FOH, and the other channel will send the same signal to your monitors. This way you have two separate controls for both FOH and monitors so that any changes you make to one won't affect the other. I carry a mini jack to phono cable with me all the time, and I have adapters that convert phono to XLR—you'll never know when you need to plug an MP3 player into an XLR socket.

Headphones

My favorite monitoring headphones are the DT770s from Beyer Dynamic. The difference between monitoring headphones and normal HiFi headphones is that HiFi headphones are acoustically designed so that the bottom end is rounder and the higher end is sweeter, for example. You should really only use monitoring-type headphones, as the coloration in Hifi style headphones won't give you a particularly good reference for the signal you are listening to.

Whatever headphones you get, make sure that they are *closed-back headphones*. Closed-back headphones have much more isolation from external noise. That makes them far better for monitoring the signals in the console, unlike open-back headphones, which transmit a lot of spill from the outside, making it very hard to listen to a specific thing in the mix. Open-back headphones have a lot more ambient noise in them because they are isolated like closed-back headphones. They are typically used in the studio, or even in the street so you don't get run over while crossing the road, but if you try and use them at a live show you might not be able to hear what you are supposed to be listening to because of the volume in the room.

CD and Microphone

You should carry a CD with a selection of well-produced songs that you can use to tune the PA system. It's also a good idea to have a mic in your case; with a reference mic you will know what your voice sounds like in case your CD player breaks down.

Notebook

Lastly, I always carry a notebook with me. It is a good idea to have your notebook out over the first couple of shows, so you can make notes if there are things you didn't notice at rehearsal or that the band didn't tell you.

Toolkit

Obviously, making sure that the tour has adequate tools to function properly is important. Below is a list of a toolbox essentials, *with just enough tools to get out of most situations*:

- Soldering iron—I use a gas-powered one in case I can't get to any power.
- Solder
- Sponge for soldering iron
- Wire strippers
- Heat shrink
- Cloth
- Long nose pliers
- Set of screwdrivers – include an electrician's screwdriver in this set as well!
- Set of Allen keys
- Fuses (assorted)
- Torch—I also have a little torch that I can strap to my head in case I need both hands. I do look like a prize plum, but it works
- Knife—I have a multitool that gets me out of most smaller situations
- Multimeter
- Power socket tester
- Sharpies (assorted colors)
- PVC tape (assorted colors)
- Spare batteries
- Adjustable spanner
- Multitool
- USB flash drive
- Temperature and humidity gauge
- Cork screw and bottle opener combo

ADAPTING TO PROBLEMS

Regardless of the size of the tour, you will, from time to time, encounter problems that were simply unforeseen. Things will go down in the middle of a show: a mic might die, or a cable gets pulled out of a socket because the guitarist has run across the stage and knocked it out. There will be times when you walk into a venue and the PA is in a shockingly bad condition, even though there was no indication in the venue specs. You may find that the faders on the desk aren't working, or the speakers have been blown. The point is that any number of problems can arise while on tour, so it's important to keep this in mind and be prepared for things to go wrong. Whatever the problem, you need to be able to deal with it quickly and efficiently. There's no point in getting stressed about the situation—this will only make it more difficult to fix. Just do what you need to do, and stay calm while you're doing it. There is nothing more gratifying than the feeling that you have helped sort out and fix a problem.

Your job as an engineer is to get the best out of what you are given, even when you are given the worst. It can be tough, but it can also be rewarding. Listening to a PA for the first time and realizing that it is in terrible condition can be scary, but don't panic—stop and take time to think about the situation. In my experience, I've had to take down entire speaker stacks because of their condition: bad time alignments, bass distribution, and the PA in the wrong position. Remember, though, that you also need to be able to work with the team that is running the show; sometimes there are unknown factors in effect—for example, perhaps the ideal height for the PA is in the sight line for the disabled platform, and therefore needs to be moved up.

Making well-informed and intelligent decisions is the best way to get the best out of any show. Listen to the more experienced people—there is a reason they are more experienced. Qualifications are one thing, but in this industry, experience counts for everything.

SECTION 2
Show Day

"Listen to the stage manager and get on stage when they tell you to. No one has time for the rock star act. None of the techs backstage care if you're David Bowie or the milkman. When you act like a jerk, they are completely unimpressed with the infantile display that you might think comes with your dubious status. They were there hours before you building the stage, and they will be there hours after you leave tearing it down. They should get your salary, and you should get theirs."

—Henry Rollins

CHAPTER 7
Load-In

Hopefully, you will have received information from your tour manager or production manager about what time you are expected to be at the venue to get your equipment out of your vehicle and into the venue. This process is called *load-in*. Depending on the size of the production, load-in could be any time of the day, or even a few days prior to the show. Think about this: just because your band's set might only be an hour, you still might have to be at the venue at 9 a.m., for a 9 p.m. stage time . . . Long old day . . . But as long as you make it a relaxed and calm day you can have fun, and the whole thing becomes worthwhile. For festivals like Glastonbury and Coachella and for truly big bands like the Rolling Stones and U2, load-in could even be a week in advance of the show. These types of bands usually tour with three complete systems and rotate them from show to show because it is impossible to set up, pack down, and move onto the next show in the given amount of time with just one set of equipment.

If your production reaches the size to be big enough to have an LD (lighting designer) carrying his or her own lights, that person will usually load-in first, along with any flooring that may need to be set up. If you're carrying all your backline, PA, and lights in the same vehicle, though, it will all be unloaded (tipped) at the same time. In this situation, staggering the access to the stage is often a good idea, just to avoid too many people running into each other. Get any flooring down first, then set up the lighting while the PA techs can start setting up the monitors, FOH, and boxes. Just be aware of what is going on around you at all times, and try to think ahead about what needs to be done—not just from your point of view, but everyone else's as well.

Know what your job involves. If you are working at FOH on a full production tour (i.e., a tour that is carrying all its own equipment, including a PA system and lights), you should be in the venue before the trucks are unloaded looking at ways to get the PA system to cover the entire audience and discussing how to hang or stack the PA. If you're the one teching the show FOH or monitors, you should be in the back of the truck letting everyone know what is and isn't going into the venue. For maximum efficiency, create a specific order in which the equipment should enter and leave the building.

You should also discuss the best position for your consoles, (We discuss the do's and don'ts of FOH console position later in Chapter 9.) In the case of positioning your monitor console, bear in mind that the lighting dimmers and guitar world will also be on or around the wings of the stage, and watch out that your console doesn't block any stage access or fire escapes. As a general rule of thumb, we place monitor world on stage left. If this isn't practical, which is sometimes the case, just make sure you have good sight lines with the entire band while they are on stage.

If you aren't on a full production tour, which is most likely the case in your early years as an audio engineer, you will probably be loading closer to the start of the show. Because of this, you need to know how long it takes to set your entire backline and get monitors and FOH ready for the band to come in and soundcheck. If you're running late, be sure to tell the tour manager; there's nothing more annoying than having a band on stage playing their instruments while you're trying to sort out some technical problem. I was working with a bunch of musicians once who really loved playing their instruments, to the point where they just couldn't put them down. This isn't normally a problem, but they would all come in early and plug in and jam before check. We ended up having to fine them their Per Diems every time they did that because we just couldn't get the stage set up in time for soundcheck, no matter how much we told them.

There are some basic unwritten ground rules about load-in and load-out that you should always keep in mind:

- You are all on the same team, so work as a team. You aren't finished until all your equipment is in the back of the tour vehicle.
- Never be on your cell phone while others are loading in (except in the case of an emergency, of course).
- Don't just watch people. If you're done, move on so others can carry on with what they're doing. There is nothing more frustrating and morale killing than just seeing someone sitting on a flight case drinking a beer watching you while you are still hard at work.

LOCAL CREW

The local crew are the people who work at the venue du jour permanently, and they're there to help you load-in, set up, and pack down. There is usually a crew boss, who's in charge of keeping everyone organized. If you have an experienced local crew, they'll normally know which flight case goes where, the different terminology for different pieces of kit, and how to set up the equipment, but they won't know your particular setup, so supervision will probably be needed. Most of them will probably either do this professionally, or do this to fill in the time between tours. Obviously, this helps with any setup. Ideally, you should split up the local crew into different teams to work with the different members of your own crew. They might have already split themselves up in to different teams because they might have a background in lights or PA, which is very useful. This will help utilize them as efficiently as possible. Always try and keep the local crew on your side, give them precise and clear instructions, but also be polite and friendly. If you have a problem, talk to the crew boss, and let him or her sort it out.

Some of the best crew I've ever worked with are in Glasgow, Scotland. These guys are amazing. I once saw four of them run a Midas XL4 up four flights of stairs at The Barrowlands. If you don't know what a Midas XL4 is, then let me just say that it weighs the same as about two fully grown space ships and is the mother of all live mixing consoles. It was seriously impressive.

CHAPTER 8
Public Address Systems

Public address (PA) is a general term that refers to all types of amplified speaker systems, whether it's a small system that has been permanently installed in a church or a system that has been installed for a day in a big venue for a five-piece rock band. The original definition of *PA system* was a sound system used for making announcements. Now the term also is slang for a musical sound-reinforcement system.

A *sound reinforcement system* is what we use to amplify our music. I think of a sound reinforcement system as a system in a small venue that just reinforces the sound coming off of a stage, so you get a mixture of on-stage sound backed up with amplified sound from your sound reinforcement system. When you start getting in to venues that are bigger and bigger, the volume of the instruments gets less, so you've stopped reinforcing the sound and are now just amplifying to make the music louder. The phrase "sound reinforcement system" is rarely used; "PA system" is much more common, most likely because it's just easier to say.

In the context of this book, we are going to specifically focus on PA systems that are used in places where bands are playing.

PA TECHNOLOGY

On August 15, 1965, the Beatles performed a show at Shea Stadium in New York City. Properly amplified PA systems didn't exist at this time, so, at most venues, the vocals were amplified through a small PA system, and the rest of the sound would just come straight from the stage. This concert, in front of 55,000 people, was no exception; the Beatles had no other choice but to use the stadium's PA system (a Tannoy type that is used just to announce that someone has parked his or her car in the wrong place). Reports from the show state that the crowd was so loud that the band couldn't be heard at all. Despite this, the concert was one of the first successful large-scale shows and paved the way for the mega bands of the 1970s.

The impetus for the development of PA systems in the 1970s was the famous acts that were around at the time. Bands switched from playing smaller venues to playing huge stadium tours. At this point, technology was limited, so the techs behind the scenes had to invent ways of getting the audio across to most of the audience; as a result, this is the era that really shaped the way the industry is today. If it weren't for bands such as Led Zeppelin, Bruce Springsteen, The Police, and Pink Floyd, there would have been no reason to develop improved audio technologies. This is the era when a lot of the major players in the audio industry popped up—companies such as Turbosound, Soundcraft, Midas, Klark Technic, and Yamaha—some of which are still around today.

PA technology is constantly advancing and changing. There is always a new product on the market. It's important to try and keep up with these ever-changing technologies. However, just because a piece of gear is new and pretty doesn't mean that it's better than what we have already. Certain technologies come in and out of fashion, so use your ears and trust your initial instinct. As we have seen so many times before, and will continue to see, we will always try to make things more complicated than they need to be just because the technology is there to do it.

Principles of PA Systems

As you may or may not be aware, there are many different types of PA systems in existence, all of which offer different qualities, different shapes, and different sounds. Some are hung in the air, and others are stacked on the ground, for example. With all these options, where do you begin when looking for your ideal PA system?

In the maze that is the world of PA, there are many factors to consider when deciding what to use, how to use it most efficiently, what sounds best, and what works best with your other equipment. Most of the time, the venue already has a PA, so you don't have any choices. However, there are a couple of fundamental things you need to know about PA systems as you start using them.

The main point behind using a PA system is to get consistent audio to every corner of a venue. Make sure you have a constant frequency spectrum across the venue, as well as from front to back, with no peaks or dropouts anywhere. This might sound obvious, but, due to the limitations of loudspeakers and physical attributes of rooms, it is surprisingly hard to achieve. The PA system also ensures that there is enough power to carry the audio though the audience, which is, of course, an essential element of any live show.

The best type of PA system would be two speakers that are capable of reproducing the entire human hearing range on either side of the stage, that react instantaneously to all the transient information held within the waveform, and that will cover the audience equally. Unfortunately, this type of system doesn't exist. We can't make a speaker that has a very flat frequency response, so we need to use different types of speakers and different types of processing to get the best possible results.

Now that we understand the ultimate goal, let's look at how PA systems work. Once you understand the concept of different types of speaker drivers, speaker boxes, and how they work together, you can then begin using this information to your advantage. All PA systems have their very own characteristics, so it's important to have as much information as possible when working with them.

Most of the time, if you read the PA specifications of most touring engineers, they'll read something like this: a full-range PA system, capable of reproducing 20 Hz–20 kHz and achieve 120 dB SPL at mix position. This is actually a bit of a problem, first because there really aren't any PA systems that are capable of reproducing all these frequencies at the same time, and second, how do you know how much power to put in the venue to give you 120 dB SPL anyway.

Let's address the frequency spectrum part first. You want to be able to reproduce as much of the frequency spectrum for as much of the time as you can, and the best way to do this is to use a full-range PA system. Most of the time you can get subs that run down to 35 Hz, and with some subs they run even lower. The hardest part of all is reproducing the high frequencies. A lot of PA systems start rolling their high frequencies off around 4–5 kHz, even though they say 18 kHz is the high frequency it can reproduce, but it doesn't tell you how loud that frequency is. So the filter kicks in below 18 kHz, causing the extreme high to be less efficient and potentially causing phase shift in the high end. So read the manufacturers' specifications!

The problem with specifying how much power you need is that there are so many factors to consider, such as how many people are going to be in the room and how big the room is; the biggest factor of all is how efficient the PA is. I think the best way to look at it is to have an average of 5 watts of total PA power per audience member, then work off the capacity of the venue. For instance, for a small pub with only 100 people, you'd put a 500-watt PA in, and if you had a 10,000-capacity arena you'd put in a 50,000-watt PA. If you have a really good, efficient PA system, then you might be looking at only 3 watts per person, but when you start getting less efficient PAs you could be looking at 8 watts per person—5 seems to be a good average.

We then have to split that total wattage over the whole PA system. As your lower frequencies require more power, you'll need to devote more of your total power to those, which will probably be up to half the total power just for the subs. When we start getting into the high, as they don't require anywhere near as much power as the subs, you can probably get away with 15% of the total overall power; then the other 35% can go on the mids. Have a look at Figure 8.1: This figure shows roughly how you can split the power up into the different elements of your three-way PA system for a normal rock show.

DISSECTING A SPEAKER

The principles behind a speaker and a moving coil microphone are the same—they both have magnets, they both have a coil of copper wire attached to some kind of diaphragm, and they both have positive and negative terminals. The way they work is very simple (see Figure 8.2): The copper coil is

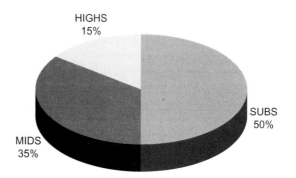

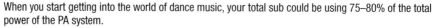

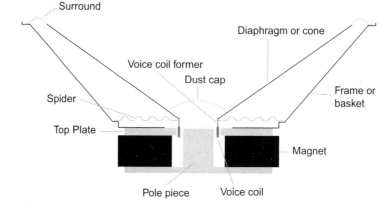

FIGURE 8.1
When you start getting into the world of dance music, your total sub could be using 75–80% of the total power of the PA system.

FIGURE 8.2
A cross section of a speaker.

suspended between the two poles of the magnet. In a microphone, when the diaphragm is moved, it causes the coil to move, which then breaks the magnetic lines of force of the magnet, which creates an electrical signal in the coil wire. Each little sound-pressure variation on the diaphragm is picked up and converted into a corresponding electrical signal. When the amplified electrical signal reaches the positive and negative terminals on the speaker, it creates a magnetic field around the speaker's voice coil, which in turn moves the voice coil in the loudspeaker, which then moves the cone in nearly the exact same way the diaphragm picked up the sound. This is how we get a replica of the same sound-pressure variations that were made on the diaphragm in the microphone, in addition to all the processing the signal undergoes on the way through the system.

Because the principles underlying speakers and moving coil microphones are exactly the same, you can actually use a speaker as a microphone and vice versa. It would sound terrible and we don't recommend it, but it can be done.

Air Impedance

Air impedance is the impedance of the air next to a speaker cone. Direct radiating drivers (which are speakers typically mounted on the front of a speaker enclosure; see Figure 8.3) are low in efficiency due to the lack of impedance matching to the air around them, whereas horn-loaded drivers (which have speakers mounted at the back of what looks like a funnel; see Figure 8.4) are very efficient. This is because, in horn-loaded enclosures, the sound is funnelled through the horn (also known as a flair), which pushes the air harder, sending it further. To see this principle in practice, try punching the air in front of you; obviously, your hand just glides through the air as if nothing is there. However, if you get a piece of drain pipe, cover one end, and then try to punch down on it, you'll really feel the air pressure inside.

We discuss speaker enclosures in more detail later in this chapter.

FIGURE 8.3
A direct radiating enclosure.

FIGURE 8.4
A horn-loaded enclosure.

Size Matters

When it comes to speakers, size does matter, though the appropriate size really depends on what kind of audio information you are trying to reproduce. Speakers must respond to the type of waveform they are trying to reproduce; for instance, high frequencies have numerous oscillations every second, so you need a speaker that can respond extremely quickly. By the same token, subfrequencies (frequencies lower than 80 Hz) require a speaker capable of handling the sheer power of what one is trying to produce.

In professional audio, loudspeakers are known as *drivers*. High-frequency (HF) drivers and compression drivers (which use a technique in which the diaphragm is larger than the hole the audio is being pushed through—hence the sound waves are compressed) are usually around .5 inch to 2 inches in diameter. Their cones are usually made from titanium; this creates a lighter, more rigid cone that responds faster to higher frequencies without breaking up into multiple vibration modes, enabling them to reproduce transients more accurately. Midrange drivers have cones that are between 6 and 12 inches in diameter.

Low-frequency (LF) drivers, also known as bass drivers, are between 15 and 18 inches in diameter. This size range makes a big difference in the sound produced; 15-inch drivers will give you more punch, whereas 18-inch drivers will give you lower frequencies. Some smaller PA systems use 12-inch cones. Sub-bass drivers (subs) are usually anywhere between 15 and 21 inches. Speakers get bigger as frequencies get lower because you need to move more air with lower frequencies.

Loudspeaker Dispersion

Another important factor in speakers is their *dispersion angles*. A speaker by itself spews out audio from all different angles, but when you put a speaker into an enclosure, you create an angle in which the audio is spread.

Dispersion angles are very important when you are installing a PA system, because you want to get the maximum coverage possible in the venue in which you are working. Dispersion angles are specified in horizontal and vertical planes; these represent the spread of the sound from the speaker box. Typically, the SPL is down 6 dB at the outer edges of the dispersion angle compared to the SPL directly in front of the speaker.

Speaker Enclosure Design

While the high-end details of speaker enclosure design are not relevant to the subject matter of this book, it is important to have a basic understanding of its principles so that you can see how it affects audio. Next we'll look at some of the ideas behind these designs.

POINT AND SHOOT

The traditional way of getting the best coverage from a PA system is to place one box on top of another, or side by side. This creates a cluster, or an array, for the audio; you can just point the system in the direction you want the sound to go, and this is why they are known as *point source* arrays. The proper name for this kind of system is *virtual point source*, as point source actually means a tiny area of space where all the energy for the sound is coming from in a spherical shape (omnidirectional). We aren't gods, and we can't make infinite amounts of power come out of the tip of a pin, so we draw our sphere and then place speakers on part of it. The idea is that when you follow the line through the back of the speakers they will all meet at a point. This would be the point source.

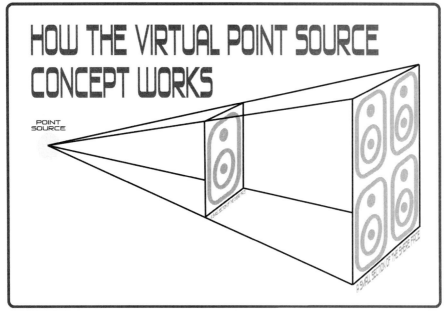

FIGURE 8.5
Shows the basic idea behind the virtual point source system principle.

In large spaces, for example, you might hang the system in the air and point it down toward the audience. These point source systems can hone in on one specific area and can be treated differently from the other speakers in the system (i.e., they can have different EQs). We place these loudspeakers so that they cover selected parts of the audience, either as a single unit or in arrays of several speakers.

In these types of point source systems, the drivers are placed quite close together at the rear of the box and then flared out at the front. Because of this arrangement, when you couple the boxes together, they create a wall of sound.

There are two typical designs for point source systems: long-throw and short-throw. Because of the dispersion angles of the speakers, we need to employ these two different types of systems so that we won't get too many points in the venue where the coverage from different speakers overlap. When overlapping occurs, this can cause all sorts of phasing problems. Long-throw systems have a much narrower dispersion angle, because you need to make the audio travel as far as possible before overlapping occurs. Short-throw systems tend to use direct radiating enclosures (not horn-loaded) that can give a cleaner sound. They are typically used in venues where the audio doesn't need to travel as far to reach the back of the room, so you need to make sure that the coverage in front of the speakers is as good as possible. When working with these systems, you need to aim the enclosures and consider the dispersion angles in order to reduce overlap of frequencies. The manufacturers will give you the measurement for the optimum gap between the enclosures and the dispersion angle of each part of the enclosure. Using that measurement combined with your eyes and ears, you should easily be able to work out where to point the boxes. Make sure you walk the room to make sure all the space is covered, and adjust the enclosures accordingly. There really doesn't need to be much science involved; just understand where the enclosure is pointing, and a bit of common sense should be all you need.

The sound level for these types of setup can be pretty big close to the drivers; the closer you are to the speakers, the louder it gets. However, the farther you are, the quieter the audio becomes. As a rough guide, as you double your distance away from the source sound outdoors, you'll lose 6 dB in volume. This doesn't really matter too much in smaller venues, but when you are looking at putting a system into an outdoor festival site, for example, the problem becomes quite apparent.

There are drawbacks to this type of system, of course. For example, it can result in *comb filtering*, which occurs when sound from two different drivers cancel each other out at certain frequencies. Comb filtering resembles a teeth comb, where phase cancellation and/or summing happens over a periodic area. In addition, flying this type of system can take time; coupling all the boxes together and hanging them from the ceiling can be a difficult task. That being said, with the right type of system and the right system tech, the results can be fantastic.

HOLD THE LINE

The other common—and, frankly, most fashionable—type of PA system is a *line array* system. The big advantage this type of system has over others is that the volume loss over distance is far less—approximately 3 dB over the doubling of distance. They are also easier to fly than conventional point source systems because they just hang in one long straight or curved line, and there is no need to worry about angles across the horizontal. All the same, you do need to make sure the dispersion angles along the vertical give you the right coverage into the room.

Some line array systems tend to use direct radiating enclosures, which inherently have low efficiency due to their lack of impedance match to the air. However, this lack of efficiency can be overcome by coupling many speakers together and by using high-power amplifiers.

Recall the statement earlier in this chapter that the best type of PA system is one large speaker system on either side of the stage. This may be a little impractical, but it isn't as far-fetched as you may think. Speaker enclosures have multiple loudspeakers, and to create an array large enough to cover a big space, you need multiple boxes. If you can get all the loudspeakers in an array to move together in unison, you can get the entire array to act as one giant speaker—or, in other words, one summed output.

Unfortunately, this is rather hard to achieve. Line array systems are designed in such as a way that all the drivers are situated close together vertically. This narrows the vertical dispersion angle while keeping a wide horizontal dispersion angle. Minimizing the separation between the drivers helps to reduce the effect of sounding like two separate sound sources. In reality, this is only partially effective. When the wavefront (the point where all the sound from the speakers comes together) is constructed, there are often phasing issues due to the way the drivers are aligned and the interaction between the enclosures. This can lead to timbral effects (very much like the effect the pinna has in the ear) that lead the brain to translate this timbral filter as space. High frequencies have a natural loss in level and transient wave shape over distance (which is one way we can tell when sound is traveling from far away). When timbral and transient information is lost in a PA system due to various phase issues or multiple arrivals of the same sound, even milliseconds apart, it makes our brain process this as distance, making the audio sound farther away. Obviously, this is not ideal at a show. It can also lead to curtain midfrequencies summing together and being more prominent than they should be, and also to total loss of high-frequency content over about 8 kHz. In addition, moving the angles between the boxes can present a different EQ, as the waveforms of the frequencies that are summing together meet in different places. In some systems, this may lead the whole system to be overprocessed, with EQ, compression, and processors monitoring what each box is outputting. In these cases, any natural sound has been lost and has been replaced with what I can only describe as a processed sound.

The wavefront that is formed seems to be fairly weak because of the wide horizontal angle. If you ever go to an outdoor event, you'll hear the higher frequencies start to disappear. This is the difference between hearing the vocal and not hearing it. And line arrays have a lack of subtlety in high-end frequencies. This is due to their use of compression drivers, which force the high frequencies out in order to help them cover more distance. They are very efficient because they match the transducer to the air load. It's not only line arrays that use compression drivers—you can find them in point source systems as well. I find this method can be extremely harsh, whether it's on a line array or a point source array. But when all the elements of the entire PA system come together, it can sound smooth.

Point Source vs. Line Arrays

In recent years there has been an ongoing argument as to which type of PA system is better. Well, as you can see, they both have their advantages and disadvantages, and you need to make your own mind up about what you want to use and how you want to use it.

Advantages of Point Source

- Point source systems come in a range of horizontal and vertical coverage angles—wide to narrow—so that you can direct the sound where it needs to go rather than spread it out across the room. Line arrays tend to create strong sound reflections off the rear wall, which muddies the sound in shallow rooms. They also radiate sound in a wide horizontal angle, which creates reflections off the side walls.
- Because the sound in long throw point source systems is more directional, the wavefront is stronger, helping it to push through environmental factors more easily. But to cover the entire audience you may need to have a lot of boxes.
- Compared to line arrays, small woofer/horn speakers can be easier to hang, typically using load-rated eyebolts and steel cables or chains. Also, woofer/horn units can be more easily mounted in special locations to adapt to the shape of the room. Line arrays require a very high ceiling because they must be tall to work properly.

Advantages of Line Arrays

- Line arrays provide more consistent volume from the front to the back of the room. The angles and the volume of each speaker box can be adjusted to make the most of the room, ensuring that the sound volume does not drop off much with distance.
- Viewed from the side, a line array radiates sound in a narrow beam, so that not much sound reflects off the ceiling back to the listeners. The narrow beam also increases the ratio of direct sound to reverberant sound.
- Line arrays are preferred for very deep or very wide spaces. They excel at projecting sound over long distances and over a wide horizontal angle as long as they are indoors.

Just like any industry, we have precision tools for doing the job, which can yield greater and better results, but when used badly they can absolutely destroy what you are trying to achieve. And just as in all industries, we have types of tools that help you do your job and achieve an all-round good sound, but it might not be the standout audioscape you want to hear every night. Ultimately, getting good results comes down to user preference and to how good you are at your job. We all want to think we are at the top of our game, but the only way to learn and achieve a better sound in the long run is to hear about the mistakes we are making and to work on them. But when we hear about the mistakes we are making, it is sometime hard to take them on board as they are: that is, as just mistakes that we can learn from.

Hard and Soft Focus

Recall that when we discussed waveforms, we saw how each wave has a rise time, or a *transient response*. Loudspeakers also have a rise time; however, because of the electromagnetic way in which loudspeakers work, and the required large size of driver cones for large venues, it's difficult to reproduce the rise time of rapidly changing waveforms. Losing that transient information causes a blurring in the audio, similar to the soft focus on a camera. This means that a proper stereo image is harder to achieve, creating space in your mix becomes harder, and ultimately it makes the audio harder to listen to. When something is in focus on a camera, the whole image stands out, it is well defined, and you can make out exactly what it is. This is the same with audio: When all the elements of the waveform are re-created, the audio become more defined, easy to distinguish, and far more easier to listen to.

Coverage, Clusters, Arrays, Infills, Outfills, and Delays

Any sort of multiple speaker arrangement is a cluster or an array. A cluster is typically a smaller array of speakers; you might have three speakers set up in the center of the stage, which would be a center cluster. An array is much bigger. The main principle of any type of array of speakers is to get the maximum amount of coverage possible for that type of speaker enclosure. Getting the right coverage from just two arrays on the left and right of the stage is sometimes going to be impossible—some venues have multiple balconies, others have extremely wide stages, and still others may not have enough space to put all the speakers in the array you need. To overcome this problem, you must use extra speakers to cover areas where the main PA doesn't reach. There are several different ways you can arrange speakers to address this problem, some of which we discuss next.

Center clusters consist of a small array of speakers situated above the front center of the stage, usually hanging from a point or a truss. These clusters are somewhat rare, normally due to the lack of points or space available above the stage. In certain situations, however, they can be done; when working with Amy Winehouse, I would feed only her vocal into the cluster, and then I would delay the rest of the PA slightly behind it. This made the vocals seem like they were right out in front of you.

Infills (front fills) are used to cover the front of the audience, which isn't covered by the PA system. They are normally situated on the sides of the stage, pointing toward the front center of the audience. Be careful when setting these up; you must avoid radiating sound toward the stage, which will result in feedback from on-stage mics.

Delays, or *delay stacks*, are situated farther back in the auditorium. As certain parts of the audio die off with distance, you use these delays to boost those frequencies. The overall effect shouldn't be noticed; in other words, the audio should sound like it has never died out and is still coming from the stage. We'll look at why we need to do this, and how it can be solved later in this chapter.

When putting a PA in a room, stick to this basic principle (regardless of what the manufacturer says): Point it to where it needs to go. There's no reason to hang an entire array from the ceiling when the audio doesn't reach the back of the room under the balcony and you don't have any delays from coverage. Also remember that if the audio is reflecting off walls, the sound isn't going to be as good as it can be—and if there is a balcony in your venue, there's going to be a lot of space where no one stands or sits and some sort of reflective surface running halfway up the middle of the venue, causing your audio to bounce back in to the room, which gives you a cloudier audio image. Use your knowledge and judgment when setting up systems: Just because the PA manufacturer or the principles of that particular type of system say that splitting the system will cause phase cancellation doesn't mean you can't do it. If done right, you can cancel the audio in a place where no one is sitting or standing. This is not a recommendation to start repositioning all PA systems—just keep in mind that the rules don't always apply, and just because someone tells you to do it one way doesn't necessarily mean it's the best way. Understanding the principles and putting a little thought into what you are doing is far better than just following the rules and creating average sound.

Subs

One of the most important things to keep in mind about subs is that they're omnidirectional, meaning that it's impossible to tell from which direction the sound is coming. All the sound appears to come from the full-range speakers, not the subwoofers.

When you fly a PA system, the subs are separated from the rest of the PA, so it's important to get a good, even spread of bass across the room. Ideally, you should make one mono source for the sub in the center of the room, but this isn't always possible (usually because of crowd barriers). If you have a left and right sub stack, make sure they are all aligned together and in the same direction so that the drivers are placed in a line together; this will help unify your sub sound across the room.

SPEAKER MANAGEMENT

After the PA is properly set up with all the correct angles and directions, the next step is to see that the whole system works together.

The Phase and Time Continuum

Correcting phase and time issues is probably the most important part of setting up any type of PA system. It involves making sure that you don't have multiple waveforms canceling each other out and giving you multiple arrivals that make audio sound very distant. When a waveform is out of time with another one, it is out of phase. You hear a different part of the cycle of each wave.

Here is another one of those industry sayings that is incorrect. When talking about time aligning, you actually align the signal, not time itself (unless you are Doctor Who). So we should say "signal alignment," but we don't.

First, we'll look at the delay stacks because it is more obvious when you hear it. Figure 8.6 shows little Mr. Stick-Man standing at the back of a room listening to some music. The PA techs have put up a delay speaker so that the sound level is still nice and big at the back of the room. Because the sound from the main

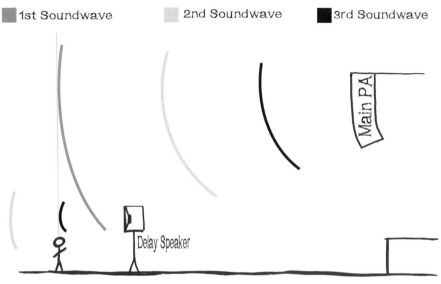

1st Soundwave   2nd Soundwave   3rd Soundwave

Main PA

Delay Speaker

FIGURE 8.6
A simple delay speaker setup.

PA and the delay speaker (which isn't yet time aligned) is being produced at the same time, the music arrives at his ears at two different times. Because one speaker enclosure is closer, Mr. Stick-Man hears the sound from the nearest speaker first. If we delay the time of when the delay speaker produces the sound, we can make all the music sound like it's in time (Figure 8.7), and Mr. Stick-Man will have a very enjoyable concert.

If there is any slight variation on an angle, or if a speaker is slightly farther back than the speaker next to it, there will be some degree of phase cancellation. This will lead to having a smeared audio image. As we mentioned before, having all your speakers aligned correctly will help, but there is also time alignment within the enclosures because if the drivers are out of alignment you will have the same smeared audio image. The goal is to have the entire speaker box completely in phase and time-aligned so that the sub, midrange, and highs create a single wavefront at the listener's ears.

When La Roux was touring the United States at the beginning of 2010, we played a show in Boston; it was a relatively small venue with about 900 capacity. While setting up the system before we soundchecked, I was walking around the room listening to what was happening with the sound in different places when I noticed a lack of punch in the system in the middle of the room. There was enough sub, so it wasn't anything to do with them. After much sidestepping and a frank discussion with the in-house soundman, we flipped the polarity switch on the low mids, and there it was. The punch was back, not just in the middle of the room, but it made the rest of the PA sit well over the entire room.

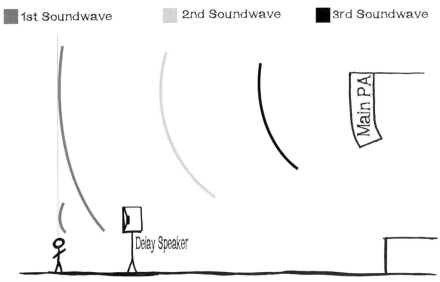

1st Soundwave 2nd Soundwave 3rd Soundwave

Main PA

Delay Speaker

FIGURE 8.7
Delaying the sound closest to you by the right amount will cause the sound from the main PA and the delay stack to seem to come only from the main PA speakers.

To fix this problem, you can apply a delay to speakers based on their location; the farther back they are, the more delay they need. You generally work in milliseconds when setting up time alignment, but you will find some systems that let you work in feet or meters. As you know, sound travels about 1,120 feet per second, depending on temperature, humidity, and air pressure. But as a general rule if you are looking at your PA and pacing things out, you want to be working on 1 millisecond per 1 foot or quarter meter. This should give you a fairly accurate alignment, but you might have to add or subtract a millisecond here or there to sink it in properly. You can use computers to help with time alignment, but computers don't have ears or the capability to understand how we hear changes in the audio; for example, you might want to have the high mids with less delay than what the computer is telling you because it makes them more prominent and produces a more pleasing stereo image.

The formula for working out what time you need is:

$$T = D/C$$

T is the required delay time in seconds,
D is the distance in feet between the main PA speakers and the delay stack, and
C is the speed of sound.

An understanding of how phase and time are linked can be used to your advantage. In smaller rooms, it's a good idea to align the PA to the backline. Delay the signal going through the PA speakers so that the speakers' sound aligns in time with the backline sound. This way, the backline sound and the loudspeaker sound are in phase with one another, and you shouldn't be getting any form of cancellation; all the sound will be arriving at the listener at the same time, which improves transient response.

At London's Shepherds Bush Empire, as far back as I can remember, there have always been a lot of low-end frequencies right through the center of the room. This is caused by the sound from two low-end drivers summing in the middle of the room. So instead of having nice consistent bass through the room, in the center there is nearly double the amount compared to just a few meters on either side. This summing starts at about 15 feet, from the dead center in front of the stage, and goes out in a slight V shape toward the back of the dance floor where the sound console is situated, for about another 15 feet. If you were to stand outside of this area, you wouldn't hear the summing of the low end at all; you couldn't even hear it in the sound booth just behind this area. There is no way at all of actually getting rid of this summing, unless you realign the PA. In this particular venue, though, realignment isn't an option; the only thing you can do is move the point at which the summing occurs from side to side by slightly delaying the signal to one of the speaker stacks. Putting on a slight delay is the same as putting something out of phase—you are moving where the cycles of the frequency meet and therefore couple, but you are able to control where the summation areas are. The reason we might want to consider moving the point where the low end sums together is that there probably will be fewer people standing in the wings of the venue, so you will affect less of the audience than you need to.

When we consider the high sensitivity of our ears, we realize that we can detect a delay of about 13 microseconds between each ear. When you start to hear multiple arrives of the same signal, our brains use that as a reference for distance, like reverb. So it's important to try and keep everything aligned as much as possible. This will keep your sound at full strength and the entire frequency range intact.

When something is out of phase, you will notice that it doesn't sound right. As we've said, your job is to get the best possible sound out of the system and room. Know the limitation of the equipment you are using and the limitations of the venue.

Crossover Points

As you know, it's impossible to build a speaker that both covers the entire audio frequency range *and* keeps the transient information needed for quality audio. To solve this problem, we split the audio frequency range into smaller ranges for each speaker, so that the bigger speakers handle the lower frequencies and the smaller speakers handle the higher frequencies. We then use what is known as a crossover to split the frequencies into different bands for each speaker. A crossover is a device that splits your signal into different frequency bands. The reason we do this is to stop unwanted frequencies from going into certain speakers, because you wouldn't want to put subfrequencies into a high-end driver, for the energy it takes to create subfrequencies would just blow the high-end driver. The crossover point is the frequency above which, or below which, the signal changes gradually from one speaker to another.

We'll talk about this subject in more detail later on, but it's worth mentioning that when EQing a system, crossover points can be a place where frequencies tend to jump out at you. When you have two sets of speakers—let's say highs and mids—there is a frequency that they are crossed over. Because each speaker is handling frequencies very close to the crossover point, we get a summing, which makes these frequencies stick out a little more than they should unless the crossover frequency response is shaped to prevent that. We use crossover curves, such as "Butterworth" or "Linkwitz-Riley," to help combat this effect. (Don't worry about the names of the curves; all you need to know is that they're variable. If one doesn't sound right, try the other one.)

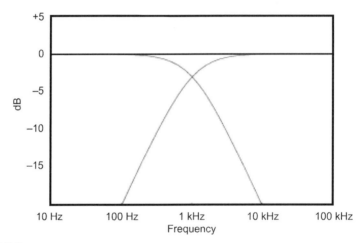

FIGURE 8.8
In a typical Linkwitz-Riley curve (or filter), the idea is to reduce the level at the crossover frequency enough that we don't hear a change in level near that frequency.

ACTIVE AND PASSIVE CROSSOVERS
You will likely encounter both active and passive crossovers in your work. Passive crossovers are typically built into speaker enclosures; they receive amp level signals and are usually fixed frequency—meaning that you can't change them. Active crossovers, on the other hand, have transistors and require AC mains power to work. They receive line level straight from the mixing console and then send line level to the amps. You are able to vary level and frequency response of all the speakers and to add signal delays as well.

You are more likely to come across passive crossovers with smaller shows. As you work bigger shows with better equipment, you'll be using active crossovers over the whole system.

Amplifiers

After coming out of the crossover, the signal hits the amplifier rack, which is where the amplification of the signal happens. An *amplifier* is an electronic device that increases the amplitude of a signal. There are many different types of amplifiers, and they have many different functions; we are only going to concentrate on the specific use of amplifiers within the live audio industry.

When working with live audio, you must match up the power of the amp with the power-handling capacity of the speakers. Typically the amplifier rms power should be twice that of the rms power handling of the speaker to prevent clipping the amp, but the amp is not turned up all the way. If a speaker is consistently fed a signal that is higher power than the speaker can handle, the speaker will eventually burn out. The rms value of the amp is an average of what it is capable of producing. The peak power (burst power) can be double that. Most amplifiers can't sustain that kind of power, and it isn't necessary anyway because the music doesn't require that amount of sustained power.

It's also important that the speaker impedance not be lower than the lowest impedance the amplifier is designed to drive, or else the amplifier can overheat. Check the amplifier data sheet to find its lowest impedance rating.

Lower frequencies require more power than higher frequencies; when looking in your amp rack, you will probably find that there is a selection of different powered amplifiers. The higher the power, the more likely it is that the amps are driving low-frequency speakers.

Amplifiers affect sound more than you might realize; for example, they can bring warmth to your low end and harshness to your top end. Because you're passing your audio signal through them, it's important that you are using a good-quality amp that works with your system; there's no point in purchasing quality L-Acoustics enclosures and then putting low-quality amplifiers in the signal path. You audio will only be as good as the weakest link in the chain.

GOING THERMAL

If you put a significant amount of signal into an amplifier, causing it to clip for an extended period of time, you run the risk of it going thermal. When it does, everything turns off and you have nothing other than a few red lights on the front of your amp—this indicates that it's gotten so hot that the contents of the amplifier will melt if it isn't turned off. This could happen for a number of reasons; for example, perhaps the fan inlet is covered in dust and no air is getting through to cool it off. Obviously, it's essential to avoid going thermal during a show. If the amps are getting hot, make sure that you have fans facing the amp rack and that nothing is blocking any airways. Open a door if you have to, just to get through the show, but ultimately you might need to be looking into the reasons why this is happening. Many amplifiers overheat if they are connected to 2-ohm loads, but a few models can handle that—check the amp's data sheet.

LISTENING

Listening to any audio that could be received as unpleasant is hard work, not only because of the clash of what you perceive to be pleasant and the sound you are listening to, but also because your brain is having trouble processing it. You could become very fatigued and irritable if it continued for any length of time. This is also the case when you are listening to something that could be overly EQed, or something that could be phasing. Our brains will try to compensate and process this information in such a way that we can listen to it without becoming fatigued or irritable, but inevitably we do. You may notice at some point that you just can't hear what you are listening to anymore. This is a sign that you are trying to process sounds that are missing something and your brain is working harder than it should to process this information. Choosing your equipment wisely and knowing the limitations that you are working in is fundamental to having great sound that is pleasant to hear.

MONITOR SYSTEMS

Before we leave the subject of PA systems, let's briefly discuss monitor systems, which are set up at the same time as PA systems. Your monitor systems are on stage, and their purpose is to allow the performers to hear themselves. There are two ways of controlling these systems: either with auxiliaries from your front of house desk or—and this is the preference of most engineers—with a dedicated on-stage monitor engineer.

The setup of a monitor system is pretty much the same as that of the front of house PA system, but instead of trying to hang an enormous array of speakers, you are concentrating on individual mixes for the performers on stage. The amount of mixes varies according to the band. If you're lucky, you might have just 3; if not, you could have as many as 18 or more.

Wedges

Wedges are single-floor monitors and can be considered zoned systems in themselves. They are usually full range, meaning that they can deliver a pretty full-frequency band, minus extreme lows and highs. I don't know any type of PA system that can reproduce the entire audible range from 20 Hz to 20 kHz.

Positioning is important with wedges. You may think that setting the wedges as close as possible to the performer is better. However, in reality, you can get better results if you set them up a little farther away. Generally, it's recommended that you set them up 4 to 5 feet (1–1.5 m) from the performer. Wedges have an optimum angle, which you can easily see just by looking at them. If the speaker is pointing toward your head, you're in the right ball park; if the speakers are pointing toward your knees, you might not be able to hear them that well.

Fills

Fills, as we saw in the front of house PA, are monitor speakers that "fill in" coverage for a small area. Singers might only want their vocals in the wedges at the front of the stage, but they still need to hear what else is happening around them. Side fills are there to give a mix across the stage for everyone. They make people on stage feel part of the show, and they can give lots of energy to the individual mixes that are set up.

Fills will normally be a full-range box with a sub for extra kick and will have a higher power than normal floor wedges, so you can get a consistent sound level over the stage. It may seem a bit excessive, but it's possible to use small line array systems as side fills. This can give you clarity on larger stages. However, make sure you're using short throw boxes; otherwise, the drivers won't sum together as they should except at the other side of the stage.

In addition to side fills, there are also drum fills, which are for the drummer (obviously). These are similar to side fills but not as powerful. You might find that they are just a normal wedge put on its end, with a sub underneath.

Time Alignment (Signal Alignment)

Time alignment can have a significant effect on the clarity of sound on stage. For example, if you find that the vocal in your center pair of wedges is unclear, there is probably a small phase alignment problem going on. Most often, time aligning will help, as will aligning the rest of your monitors to the same point. Remember, you must align audio to the furthest point of sound; usually the side fills will be the furthest sound source away, and center wedges are your nearest point. Pace out the distance between the center wedges and the side fills, and use the 1 millisecond rule we discussed earlier. Remember: To do this successfully, you need to make sure that your side fills are the same distance apart from your center wedges.

IEM (In-Ear Monitors)

In-ear monitors (IEMs) are a godsend to front of house engineers; they allow performers to hear audio clearly while eliminating unwanted wedge bleed into the audience at the same time. IEMs are basically radio receivers attached to quality earphones that have been molded specifically for the ears of the recipient, and it's the molding that cuts out all the unwanted noise. No one else will be able to use them because they won't fit in their ears. A signal is sent from the monitor console to a transmitter, which then transmits to a pack attached to the recipient (usually strapped to a belt, or even taped to the back of a dress). The signal can be either stereo or mono mixes, and is completely individual. If a band has been used to wedges, IEMs do take some getting used to—but a little persistence can go a long way. When using IEMs, you may need to add ambient mics to the IEM mix to give the feeling of space (unless you already have a lot of mics on stage), or you can now get an ambient filter in the molds of your IEMs. This feature acts like the ambient mic does by letting a certain amount of external noise through. By changing the filter, you are able to have different amounts of external noise. You are also going to have to use reverb because using a dry vocal can be difficult to listen to (we aren't used to listening to dry sounds).

The earphones themselves come in two types: molds and generics. With molds, you have impressions of your ears taken, and the earphone driver is placed inside the mold made from that impression. Generics are either sponge or rubber. They are placed over the earphone—then, if it is a sponge, it is rolled up and placed in the ear to expand; if it is rubber, it's simply pushed into the ear. The idea behind all types of earphones is to create isolation between the sound inside and outside the ear.

A parting piece of advice on IEMs: Change the batteries after every soundcheck.

EQ vs. Volume

Let's briefly discuss EQ versus volume in relation to mixing monitors. We will go into this matter in more detail when talking about putting your mix together, but for now let's focus on monitors and monitor mixes.

When mixing on wedges, you'll sometimes find that certain members of the band want to hear themselves better, and thus ask for more and more volume. This can cause problems for the others on stage and can turn into a never-ending cycle of ever-increasing volume. If someone needs to hear himself or herself better, the trick could be to make the sound clearer, not louder—this is done by EQ. Your monitor system will have graphic equalizers over each output going to each mix, and will also have a parametric EQ over each of the channels. These graphics have 32 volume controls.

Each instrument has its own frequency range, so rather than turning things up, you can just EQ other things out. This makes space for the instrument or vocal of that particular band member to be heard much more clearly. You must also remember that filters are a great way to keep out unwanted frequencies from your mix. Your high-hat channel does not need to have frequencies below 400–800 Hz in it, so why leave them in? They will just cloud up your mix.

The same is true for in-ear monitors, but you won't normally have a graphic EQ on each mix output. When being asked to turn something up, just think about where it's sitting. Can you hear it? Is it loud enough and just unclear? It's all a matter of thinking about what you're listening to and about how you can clear it up. A good tip is to remember to put filters on all effects returns to your ears. Most of the time the lows in your effects can create the same cloudy effect that is affecting the clarity of what you are listening to.

CONCLUSION

All in all, we are limited by technology—but it's still important to occasionally try and push this limit, as this is what advances our understanding of how we can build better equipment to do our jobs. We can't produce speakers and speaker boxes versatile enough to make every venue and every acoustic situation sound amazing, but we do have speakers that work well together, and we now know enough to make sure we get decent coverage over most areas. But all is still not perfect; live environments are all about compromise, but it's where and how you compromise that is important.

There are so many good manufacturers of PA systems that it can sometimes be difficult to decide which is better than the other. My advice is to stick to what you know. If you're willing to experiment, it's still best to rely on the main manufacturers. Taking risks can be good as long as you make well-educated decisions and you understand the consequences if they fail.

> When La Roux played the Corn Exchange in Cambridge, UK we put the Funktion One Res5 system in with their twin 21-inch subs (so efficient and sound so good). The last time I was in this venue we put the subs on the floor, but due to a change in the health and safety law they now must go on the stage. Of course, the stage was hollow and not fixed to anything solid. Having large amounts of sub on stage caused a couple of problems with the band, the first one being that it's hard for them to hear the clarity in the beats, and the second one is that the vibrations cause the electronic drum pads to trigger randomly!
>
> Usually you would lay these subs flat so you get a nice wide dispersion along the floor of the room, and would probably stack two or three high. In this case, because of the hollow stage, I was worried that laying them flat would soak up a lot of the energy and cause a lot of vibrations on stage. This would mean that the low frequencies wouldn't be working as efficiently and we would need to turn them up to get the right amount of sub through the room and up to the top of the balcony. Turning the subs up would then cause more problems on stage, a vicious circle…
>
> As a solution, I decided to stack two subs on their sides on either side of the stage. None of us had really done this configuration before and had no idea how it would sound. Aaron (my PA tech) told me that it wouldn't work and we'd get all sorts of problems. The subs were horn-loaded so that the flair pointed up into the room rather than across it. And it worked: full deep bass evenly spread across the room and all the way to the back of the balcony.
>
> There were also slight line issues with the amount of space all the subs were taking up on the stage, so this solution also helped with that.
>
> In this situation, taking a risk paid off—but you must remember that risks go the other way too.

My own personal preference is a point source PA system like the Funktion One series, which is a very transparent sounding PA; you can hear absolutely everything on it. It also lets the sound out of the boxes in such a way that the stereo imaging is some of the best I've heard. (Of course, the sound is only going to be as good as the weakest link in the whole system, so having an inferior desk won't help.) The disadvantage of the Funktion One series is that they can be quite hard to mix on because it is so clean and you can hear everything you are doing on it; engineers used to mixing on PA systems that cover up small mistakes find it especially difficult. These are the only PA systems that I want to use: L-Acoustics, D&B, Funktion One, Meyer, EAW. I have a great show pretty much all the time on these. But when on smaller tours you don't always get the choice.

CHAPTER 9
Desks Up!

The next thing we do after we have raised our PA system toward the heavens is to set up the mixing consoles. In this chapter, we cover mix positions, mixing consoles, and outboard gear.

MIX POSITION

Life on the road can be hard, but it's a lot harder when you can't hear what you're doing. Unfortunately, there are many venues where the mixing desk is not situated in an ideal location, such as in a room at the back of a club, or under a balcony where you can't see the PA and can hardly see the stage (this happened to me at a show in Chicago while out with Amy Winehouse). Unless you're carrying your own production, though, there isn't really too much you can do about it. Especially if it is a seated venue, where taking up expensive seats is not in the promoter's or tour manager's interest, and the artist wants the great sound that you give them, and also the money... it's a complex game.

From time to time you might come across a mix position that will cause you problems, and you know it just by looking at it. Then you hear someone say, "You can see the speakers, so you can mix from there." This is just not cool. Of course it's always nice to be able to see the speaker you are mixing from, but it doesn't work if you are in a sound booth at the back of a club with a piece of glass in front of you. That is classed as another room, and mixing in another room from the band is like driving your car at night on a dark road. You have your headlights on, but sorry officer I didn't see the drunken man running out into the road because it was dark.

So what is the sweet spot of a mix position if you have the opportunity to be able to place your console where you want it? Each room is different, and each PA has its own characteristics; what might be great in one venue might not be great in another. Always remember that you will probably have to compromise at some point and keep other factors in mind. For example, while placing the mixing console right in front of the balcony might seem like a good idea from an audio perspective, it might not seem like such a good idea at the end of the night when your lovely new console is covered in beer, food, or, even worse, a popular sugary carbonated drink (this actually eats away at the circuits inside the console) that have been rained down by the unknowing audience above.

If you want to mix for the audience, you need to become part of the audience . . . but with a little more space than everyone else.

It may seem like I've painted a very bleak picture of live shows here, but it really isn't that bad. Most of the time everyone does care and wants to help you out, but sometimes you need to be able to compromise for the sake of the audience, promoter, and/or the tour manager.

Let's have a look at some typical positions in which you could place a mixing console.

Positions to Avoid

One of the most important things you can know about setting up mixing consoles is the positions to avoid. And if you do have to set up in these positions, you should understand exactly what you're getting yourself into.

First, the worst place in the world to set up a front of house console is in another room—that is, in a sound booth. Venues sometimes use these in the interest of security; if everything's together in one room, it's easy to lock it all away. (And, again, you're not taking up any valuable audience space.) As much as this might be an advantage for the venue and promoter, you'd do just as well mixing the show from the pub around the corner. No matter how big the hatch or window to the room is, the sound is always massively different from what the audience hears.

Another place that should be avoided is under balconies—or, more accurately, anywhere near balconies. If you're immediately below the balcony, you run the risk of people spilling drinks or food on your equipment (as we mentioned). However, if you're under the balcony, you get reflected frequencies from the ceiling, which can cloud your audio image and give a false impression of the balance in the room. Another risk of being under the balcony is that you may not be able to see the PA, which means you aren't getting any direct sound from the PA, and thus you're trying to mix something you can't hear. Again, you might as well go to the pub around the corner.

When it comes to audio, walls are usually pretty reflective. Being pushed into a corner or up against the back wall of a gig will give you a colored image of the sound. Low-end frequencies are a lot more noticeable within these areas. When you come within about 9 feet of a wall, you are beginning to enter what is known as a *pressure zone*, or *boundary effect*. The overall volume or possibly certain frequencies are increased in this area, which happens because of standing waves. When soundwaves strike a hard surface, the waves are reflected and combine with the incoming waves, so the sound pressure is increased near the surface. These reflections also cause comb filtering and boomy bass, which lead to very undefined audio and can make it generally impossible to accurately mix a show.

The other no-go area is on the side of the stage. This location is perfect for a monitor engineer, but front of house engineers won't be able to hear what they need to hear because they'll be behind the speakers so all the audio is heading away from them.

Acceptable Mix Positions

One example of an acceptable mix position is on top of a riser. If you're going to be on any sort of raised platform, though, you need to understand a few principles. Sometimes above the audience there is what is known as a *humidity layer*, which is a layer of heat and moisture generated by the audience sits a few feet above the heads of the audience, very much like the floor of a rain forest. This layer acts like a shield and causes some of the harsh frequencies to bounce off the top of this layer straight into your hearing field. As a result, you hear a lot more high-end frequencies than you would hear if you were in the audience. The effect varies depending on the size of the venue and on whether you are indoors or outdoors; heat and humidity are much more likely to be trapped in a small indoor venue than a large outdoor festival. If you do end up on a riser, make sure to get off it and listen from the floor every once in a while.

The side of the room is another acceptable place for the mixing console. You won't be affected as much by standing waves; you will, however, still be getting reflections off the walls, and you should be careful you aren't too near the pressure zone. The obvious flaw with being off to one side is, of course, if you use a lot of stereo imaging (such as panning guitars left and right) or effects that are in stereo; if this is the case, you won't be able to hear the stereo imaging that is happening elsewhere in the room. Center-panned images will appear to you to be all the way toward the side on which you are standing.

Ideal Mix Positions

The ideal mix position is the center of the room. This is the sweet spot where all points of audio come together and are audibly in focus. If your system is set up correctly, you shouldn't have any summing down the center line, which is caused when two identical soundwaves meet and then sum together, causing certain frequencies to be louder. As you are mixing the show, it's always a good idea to move left and right on and off the center axis; sometimes you can't get the system to sound completely perfect in the center, and being slightly off center can help. Remember, though, that summing can still occur for the same reasons, and when you are off center you can get some forms of cancellation. Remember that you are mixing for the audience; moving on- and off-axis will give you a better indication of what the audience is hearing.

As should be clear by now, mix position is just as important as speaker placement. Despite this, it isn't always possible to get an ideal mix position in a venue—you will have to compromise from time to time.

MIXING CONSOLES

The mixing console is the link between the artists' music and their audience, as well as the link between the engineer and the audience. It is where all of the manipulation of individual instruments and sounds happens. The first time you look at a mix console, you'll probably find it a bit daunting, but don't worry; once you get familiar with it, you'll be comfortable in no time.

With all mixing consoles, there are two main sections: the input section and the output section. The input section is made of several repeated vertical channel strips. Each strip has controls running from top to bottom, and each strip controls the sound of one microphone, DI, or other source. The output section is a little different, but follows the same idea by giving you control over all your outputs: your main left and right output, and sometimes a mono output (next to the L/R fader/faders). It will also sometimes have VCA (Voltage Control Amplifiers) faders (not all desks have these), and a subgroup section. The auxiliary send master outputs and auxiliary return master inputs are set from here as well, along with any matrix sends or two-track outputs.

Analog and Digital Mixing Consoles

The world of analog and digital mixing consoles collided some time ago. Yamaha released the DMP7, which was an 8-channel digital mixer, in 1986; however, the technology wasn't yet there to support digital consoles. We had to wait until 1998 when Harrison and Showco teamed up to create the LPC console and then in 2001 when Soundcraft introduced their Broadway console. Both of these consoles were digitally controlled analog consoles, which meant that they were still analog consoles but they had the ability to recall. By this time, technology had advanced enough to allow manufacturers to start introducing digital consoles, but they still weren't quite rivals to analog. In 2001, Yamaha introduced its PM1D, which was really the first digital console to catch the eye of engineers. It had an external rack unit that was essentially its brain, and the user surface was large enough so you didn't have to flick through layers and layers of menus to get to where you needed to go. But most important, the cost of the console was a mere fraction of the cost of other consoles of that type. Then, in 2004, Yamaha introduced the PM5D, which became a massive hit for monitor engineers. The age of the digital console had well and truly begun.

LAYOUT

The layout for all analog consoles is pretty much the same because the channel strips are laid out the same way. This has pretty much to do with the way the signal flows through the channel strip. There are some exceptions, of course; the Yamaha PM4000 springs to mind because the pan pots are at the very top of the console, along with all your routing buttons. In addition, old Hill Audio desks are pretty different from the conventional layout used now. (If you do come across one of these, treat it carefully—it should be in a museum.) But by and large, all analog consoles have the same layout.

However, this is not at all the case with digital consoles; rather, they all seem to have a *different* layout. Just like the first pioneers of live mixing consoles, everyone has their own way of doing things, so no one has yet adopted a universal format—and to be honest, I'm not sure they will. If you look at something like the PM5D by Yamaha, the audio quality is subpar, but the layout works really well for monitor engineers. I took Digidesign's SC48 out on a tour with me, and, as a front of house engineer, I thought it was the best layout Digi had come up with yet, but the audio quality still isn't what I'm looking for.

DIGITAL TERMINOLOGY

Let's have a look at this terminology and demystify what the terms are referring to and how they are related to each other.

Bit Depth

Bit depth is how accurately the waveform is measured during each sample and converted into digital data. Bit depth (word length) refers to the number of binary digits that are generated during each sample, such as 8 bit, 16 bit, or 24 bit. In a general term, this gives us the amount of binary digits available to store data. For example, an 8-bit code has 256 storage spaces, a 16-bit code 65,536 storage spaces, and a 24-bit code 4,294,967,296 storage spaces. In terms of colors, these are the amounts of different colors available in your color palette; thus the higher the bit depth, the more shades of color are available, and the transition through each shade becomes better and better. In terms of audio, this gives us a larger dynamic range; this means that the higher the bit depth, the more audio we can store before we run out of headroom and get distortion.

The best way to look at how the bit depth will affect the audio is to visualize it. Take a look at Figures 9.1, 9.2, and 9.3. These show the difference in quality between something that has a bit depth of 4, 8, and 24, respectively.

If we translate this to audio, you can see, or rather hear, the difference in the bit depth.

The higher the bit depth of each sample, the less the distortion and noise the audio has, and the more defined it becomes; with a higher bit depth we have a more accurate measurement of the sine wave voltage at the instant it is measured or sampled.

Sample Rate

Sample rates are the amount of measurements of audio per second, such as 44.1 kHz or 96 kHz. This is the high-fidelity reproduction of your sound. Think of a film: To see a continuous moving image, we need a frame rate of about 24–28 frames per second. This is exactly what the sample rate is, but more of an audio image. Higher sample rates will give you higher frequencies, whereas lower sample rates cut out highs, making the audio seem dull and lifeless.

FIGURE 9.1
4-Bit. As you can see by this wonderful picture I took on a driving holiday through Monument Valley, Utah, the colors don't blend into each other very well at all; they're grainy and not very well defined. To see the color version of this image, visit the companion site at www.liveaudiobook.com.

FIGURE 9.2
8-Bit. As we increase the bit depth the more colors become available, so we can see more definition, but there is still not that much definition between different shades of the same color. To see the color version of this image, visit the companion site at www.liveaudiobook.com.

FIGURE 9.3
24-Bit. Now we have increased the detail a lot more. The detail is far more defined, and the colors blend into each other much better than the other two images. The image is also a lot brighter. To see the color version of this image, visit the companion site at www.liveaudiobook.com.

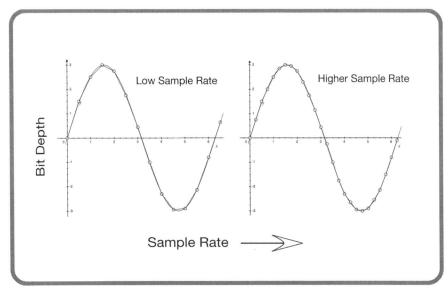

FIGURE 9.4
In this image we can see how a continuous curve of a sine wave is measured (sampled) thousands of times a second to result in *digital numbers*. The higher the sample rate, the more accurate the measurement will be, and the better the quality of the digital audio becomes.

The rule is that we must sample at twice the maximum frequency being sampled; the threshold of human hearing is 20 kHz, so this is why we get CD quality audio rated at 44.1 k, which is 44,100 samples or measurements of audio per second. Professional audio works around 96 k per second, and we are now starting to see the introduction of 192 k.

WORKING TOGETHER

When we put our sample rate and our bit depth together, we get the overall sound quality. The higher we can have our bit depth and our sample rate means the more accurately we can sample and then reproduce the sound. But there is a limit to how accurately the analog waveform is sampled and then reproduced at the other end. We sample a sound and then reproduce it; the audio processor (digital signaling processing—DSP) looks at the audio and gives it its best guess. You can't actually hear the digital process, but digital has a very distinct sound, which probably has to do with the different filters inside the audio processors.

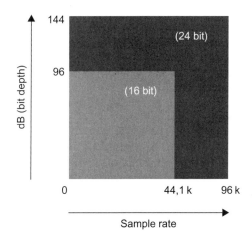

FIGURE 9.5
This little graph shows basically the amount of information difference between lower and higher bit depths and sample rates.

Bit Rate

Bit rate (also referred to as *data rate*) is something else entirely different and shouldn't be confused with a sample rate. A bit is a unit of computer processing, and a bit rate is the amount of bits that is processed within a second. Bit rate does not affect file size; it affects how rapidly a digital signal transfers from one component to another.

Bit rate is bits per sample x samples per second—in other words, bit depth x sample rate. So bit rate is a measure of audio quality, combining bit depth (which affects noise and distortion) and sample rate (which affects the high-frequency response).

Digital Signal Processing (DSP)

Digital signal processing (DSP) is the brain behind your console and is the reason digital consoles work. It gives your console the power it needs to perform all the processing that you need to run your show. Within digital data, there is a certain amount of error correction; this is controlled by the DSP.

Digital Connections

Using digital connections is important if you are going and staying digital; they keep noise and phase shift out of your signal path by avoiding unnecessary D/A and A/D converters. There are a number of different types of digital connections for connecting up your console to various pieces of kit, the most popular of which are AES and S/PDIF. AES is the professional way of sending and receiving a digital signal; it sends a digital signal to a crossover or an amp, keeping the entire signal balanced and in digital format until the last possible moment. S/PDIF also keeps the signal in digital format until the last possible movement, but it is unbalanced. These connections seem to offer the best kind of digital connectivity.

Latency

Unlike analog equipment, digital equipment takes some time to process information—and the more plugins or information that requires processing, the longer it takes. Because each signal is being delayed by different amounts due to different processes over different signal paths, the signals become out of sync. To fix this problem, the DSP must delay everything slightly, until all the processing is done, so that all the signals line up together. The time could be up to 10 milliseconds and is referred to as the *latency*.

DIGITAL PROCESS

We need to convert our analog signal into ones and zeros for our digital consoles to process. On our input stage we use analog to digital converters (A/D or ADC), and on our output stage we use digital to analog converters (D/A or DAC).

A sine wave in an analog circuit is one continuous curve, but is sampled into numbers in a digital circuit; this is our sample rate. The more samples, the finer the steps are between points of the waveform, meaning that the waveform is captured more accurately.

The digital signal is then converted back into an analog signal using a digital to analog converter (A/D). The binary numbers contained in the digital signal correspond to a specific place on the waveform, and a voltage from the A/D converter is used to reproduce the analog waveform as accurately as possible.

It's in these processes that we can get a distinct sound, and it all depends on how these converts work and are programmed. I might say that it sounds digital, and to me it does because that word for me sums up the sound, but other people might use the words "crisp" or "clean". But each converter sounds different, and it's up to you to find out which one you like. Just use your ears.

ANALOG VS DIGITAL

The hard question, of course, is whether analog consoles are better than digital ones. I personally love the sound of analog consoles, and I probably always will. However, that being said, I have used the Midas Pro6 (a digital console), and think it does an amazing job. The EQ section is extremely accurate, and the gain section responds like an analog board. For all intents and purposes, it is sonically superb—and dare I even say it, it might actually be the best console I've ever heard.

The answer to which is better, analog or digital, really depends on what you want to do with it. Are you looking for a very clean, precision sound, or are you after a grittier, or warmer sound. It's about selecting the right product for the job.

The difference between the analog and digital is like the difference between 35 mm film and high definition (HD)—both are superb in their own right. The film, which some people class to be warmer, has a certain quality to it that just looks good. But then the quality and the detail involved with HD can be utterly breathtaking. If we go to a cinema and watch the same film in HD and film, we should have the best picture quality available. There will be a difference, but they will probably come into the realms of taste, rather than one actually being better than the other. If we watch the same movie at home on an

FIGURE 9.6
The Midas Pro6 Console at London's Brixton Academy.

old TV using a terrestrial broadcast, the quality will be nowhere near as good. The sound quality of both analog and digital consoles comes down to the components that are used and the circuits that are designed. It's all about the quality of the equipment that is used through the entire audio path; everything can suffer from poor quality components.

I was working at a Brit Awards launch party in January 2010. It was held in London's IndigO2, which is in the O2 arena, which had a JBL Vertec PA and a Soundcraft Vi6 console. I've come across this combination of console and PA before, and I've always thought I've heard actual processing—not dynamic processing, but the actual CPU crunching digits. I know the processing is happening around 96,000 times per second, which apparently we can't hear; it could be phase shift in the digital filters, but it's the only way I can describe the sound. I'd thought I'd heard this before with the same combination of PA and console, at a different venue, and no one else could hear it. At the time, I concluded that I was going crazy and that my suspicion of digital equipment was clouding my judgment. However, when I walked into the IndigO2 and heard the same thing, I knew it wasn't me. The system was put into the venue by a different person, had been set up by a different person, and was being operated by a different person—yet the results were the same. And I've heard the same thing over and over again as this combination has started to be put into venues all over the world. The point is that different products bring out different effects in other products. Combining two elements that bring out flaws in each other is difficult to cover up, and it is one thing I've noticed happening more with digital consoles due to the precision of the controls.

Remember how we looked at hard and soft focus in relation to PA systems? Well, it's the same here. Analog equipment sometimes has more of a soft focus; it rounds things off and can make our audio sound more natural. Digital equipment can be more hard focus, which means the quality of the audio can be extremely good. Because the digital process has quantifiable numbers, it can be extremely precise, whereas analog works on a more ballpark principle—not to say this is bad by any means.

Obviously, as budding live audio engineers, we need a console to sound as good as it possibly can, and a lot of the sound of the console comes from the preamp. With great quality preamps you can get a great sound. Unfortunately, I haven't heard a digital preamp that I like yet, and this has a lot to do with the filters in the A/D conversation. There are also some pretty ugly-sounding analog preamps out there too, but you can never go wrong with a Midas.

The biggest problem with digital consoles that comes up time and time again is the layout, causing you to have to relearn thought processes each time you get behind a desk. Think about driving a car. When you first learned to drive, everything was coming at you so quickly that you wondered how you could ever control such a machine, but then, you got used to it. Each time you approach a stop sign you know what to do, and the more you do it, the less you have to think about it until the thought process becomes instinct. With a digital console, it would be like getting into a car and the pedals being reversed, and the indicators and windscreen wipers have been switched around. When you approach the same stop sign, you have to think about what you are doing, otherwise if you were going to turn left, you'd put the windshield wipers on and end up in the front yard of the building opposite.

For me, the digital layout, which requires various combinations of button pushing and multiple menus, has changed the process from being natural and instinctual to requiring a lot more thinking. This is because of the amount of different layouts we come across. With analog consoles the layouts were pretty much the same, so grouping thought processes together was easy. But when you are presented with a different layout every time you step up to a console, it's much harder to do. That might force us to be less creative now than we used to be, even though the tools we have might be more precise.

In addition to reducing creativity, the complex layout in combination with the ability to save has led some engineers to become lazy. For example, you may hear something you want to change, but by the time you navigate the layout to get where you want to be, the moment has passed. So rather than adjust and then readjust, you just might not bother. Similarly, when you load up your settings from the night before, you may think that they are all right for the room you are in that day as well, and thus you might not bother to change anything. Of course, the upside to this is that if you aren't able to have a soundcheck, you still have a very good starting point for the show.

The biggest advantage that digital has over analog, though, is the ability to save your settings. No longer are you required to spend ages after soundcheck marking down all your fader and pot positions, and going through all your outboard doing the same thing. Even more importantly, you are able to load the exact settings before your show, rather than translating what you have written down after soundcheck. But the major disadvantage related to this method is that most consoles use their own type of programming language. There are a couple of systems in which you can load up show files from a different console, but only within the same family of consoles; you still need to have different files for different systems. It would be great to have one type of programming language for all consoles so we could transfer files in between each console, but I can't see the manufacturers going for this. You could get the best sound you could on a Midas console and then transfer it to a Yamaha. You wouldn't actually get the same sound because of the fundamental differences between the audio ethics of the companies. For example, Midas says that "you can overdrive the input on a console," whereas Yamaha consoles are more along the lines of "you shouldn't overdrive the input because that isn't the correct way of doing things." Besides these fundamental differences between the manufacturers, I'm sure there must be a way of integrating files into one another. Obviously, each console sounds different, and there will be variations between the sounds of the boards. However, 2.5 k is still 2.5 k no matter which way you look at it. Even if a particular console doesn't have that value in its filing system, I'm sure the nearest guess would be sufficient.

Even within individual consoles the way to store data is very complicated because of the different types of filing systems they use. The files have different types of *layers* within them that are responsible for storing different parts of the data, and because of the way they store this data, you aren't able to store your show and then recall only part of it. This comes into play when you think of a festival environment. When you turn up to the FOH console with your show file and load it into the console there, chances are you will erase all the output data the festival techs have been working on because you must load your whole file. If you had the ability to only recall your channels, EQs, dynamics, and VCAs without affecting your input and output patches, masters or matrixes, we'd be happy. This has obviously been taken into consideration by a lot of manufacturers because we are starting to see a few work rounds to these problems. Some consoles have the ability to "safe" sections that aren't affected by scene changes, or, in certain cases, different show files. We are also now seeing the ability to store certain sections of the console into a "preset" folder, where you are able to save a patch, or an output section, load up your file, and then recall a preset. This is leading to a much more versatile system and a universal understanding for storing and loading show files.

The other major difference between analog and digital consoles is the visuals. With analog consoles, you only have the channel strips to refer to, but with digital you have brightly lit screens. On some digital consoles, the screen placement is poor; that is, it's stuck right in front of your face and can be very distracting. More than a few engineers have gotten caught playing around with plugins rather than mixing the show. In addition, because there is a frequency graph of what EQ you have over each channel, you may sometimes find yourself *looking* at what you're hearing, rather than *listening* to what you are hearing.

A few years ago, I was in Aberdeen working with the Welsh rap group Goldie Lookin Chain. I was at the other end of the multicore working the monitor system, while the FOH engineer was using the house console (a Midas H3000). Everything was going well, until, about halfway through the show, a pint of beer landed on the console. If this were a digital console, we would have had a problem, but, because we were using an analog system, we were able to save the situation. With the channels we were using covered in beer, we pulled out the XLRs in the back of the console and plugged them into the channels that weren't wet—a true gig saver. Meanwhile, while the show continued, the channel strips were unscrewed and pulled out of the console and dried as much as possible with paper towels (to reduce the damage of the beer coating the circuit board and eroding the circuit). Some switch cleaner and a good scrub later, and the channels are ready to be used again. This is not possible on a digital console because individual channels can't be removed

Digital consoles don't have much audio traveling through them; they just manipulate the audio that goes into the external signal processors; even if the console has inputs on the back of the console, the audio still isn't traveling through the amount of circuits it does in an analog console. Also, because the D/A and A/D converters are only manipulating data, and not actual signal, they don't add any noise to the signal path (as analog consoles do), unless you are adding an external piece of analog equipment. Most of the time, a digital console will have a digital multicore, which means that the analog audio can stay on stage. Thus, your console is just a user surface that controls the digitized audio, whereas the analog system sends the audio out to the console, sends the signal through the console and whatever outboard you are using, and then sends it back down to the amps. There is a lot more cable involved when doing it the analog way, so you are adding not only noise, but resistance from the cable as well.

The funny thing is that with all the advantages you get with digital consoles, they are also the fundamental problems with them as well. You wouldn't find any of these features on an analog console (things like the scene scope, save, and the undo button). Digital consoles might be great, but they haven't half made life as an engineer more complicated.

Underlying all of these advantages and disadvantages are the cost factors associated with consoles. Everything comes down to budgets, and digital consoles offer far greater versatility, a much smaller footprint, and a far cheaper price tag. Having dynamics on each channel, as well as internal graphic equalizers and effects, cuts the cost of needing extra racks with these units in them.

THE CHANNEL STRIP
Preamp Section

A mic level signal traveling through cables is so low in voltage that it needs to be turned into a line level signal when it reaches the console. The physical position of the gain pot (which is the physical knob that you turn) is usually at the very top of the console (unless you are using an old Hill Audio desk, or maybe a Yamaha PM4000). The gain pot itself can be called by other names, such as trim, or you might even see the letters HA next to the pot, which stand for *head amp*, which all do the same thing: They control the gain of the mic preamp.

You want to be able to control your signal level so that it is loud enough to hear, but not so loud that it clips and distorts. When a signal clips, it is hitting the end stops where there is no more room for any more signal and it starts to flatten part of the waveform. This is what causes the distortion. With analog consoles, clipping can be a pleasant experience; it can add a rawness that might be the extra push your mix needs. Digital clipping, on the other hand, is not very pleasant. As we mentioned before, when we discussed dB FS, there are only a certain number of bits in a digital signal, and this leads to a harsh crack when the signal voltage exceeds that represented by the maximum number of bits available.

Any analog-to-digital conversion follows the preamps, and, for this reason, the pre-amp and gain trim are probably two of the most important parts of the console.

+48V

In the preamp section, there is also a +48v button, which is the Phantom Power on/off switch. If you are using any form of powered microphone (a condenser mic), it requires phantom power and is most likely turned on by this button. On smaller consoles, you might find this button on the back of the console next to the XLR input, or there may be a group phantom control. Sometimes there is an overall Phantom Power control.

DIRECT OUTPUTS

Direct outputs are outputs direct from the preamp; they are not affected by any EQ, fader, aux send, or anything else on the channel strip. Some consoles let you set the direct outs to pre-fader or post-fader, so be sure to set them to pre-fader when making a recording. They are mainly used to make a multitrack recording of whatever the input is to that channel, or side chaining a compressor. You may find that some consoles have a direct-out level control so that you can control the output level, just like having a fader. This is very handy, but not all desks have them.

PHASE SWITCH (POLARITY SWITCH)

You can use this button to switch the polarity of the signal. That is the same as reversing the connections to pins 2 and 3 in the mic's XLR connector. You might find, for example, that having two mics on a guitar amp causes the whole thing to be out of phase (or opposite polarity), due to either an incorrectly wired cable or a misplaced mic. Either way, pop this switch in and hear the difference.

You will also need to use this switch when micing a snare drum from the top and the bottom. This is because, when the snare is hit, the skin on both the top and the bottom bends in the same direction—when the drum stick hits the top skin of the snare, it causes the skin to move down, so the mic picks up the waveform being created by the snare as a downward motion. At the same time the skin on the bottom of the snare is moving in a downward motion, but because the mic is pointing up toward the skin, the waveform the mic picks up is opposite to the top mic. This will probably mean they're out of phase when put through the PA. Hit the phase switch, and hear the drum come back to life.

EQ Section

These days, desks have what is known as a *parametric EQ*. This type of EQ is sweepable, meaning that you have the ability to change the frequency you would like to adjust. A parametric EQ is any EQ that gives you control over the three main characteristics of spectrum manipulation, notably:

- Frequency: You can select the frequency you want to control.
- Amplitude: A gain pot for the frequency you have selected. You can apply cut or boost of that frequency.

FIGURE 9.7
The channel strip.

■ Bandwidth: Normally written on the console as the letter "Q," which stands for "quality," this setting controls the bandwidth—the amount of frequencies on either side of the selected frequency that are affected by the cut or boost of the frequency gain control.

Q is really important when controlling the amount of bandwidth you want to reduce or increase in volume. For certain things, like vocals, you might want to have a wide bandwidth (low Q, like 0.5 to 1) at the frequency of 1.5 khz and only pull it back a dB or two, just to remove some of the honkiness that you get from that area. However, this might cause some sibilance at 8 kHz, and you don't want to pull out frequencies there, as it might make your vocal sound a little dull. In this case, grab your Q knob, crank it to the smallest bandwidth (high Q, like 5 to 10), and then cut as much as needed without affecting too many frequencies around the one you need to take out. You can also put the Q on the smallest bandwidth setting and then boost the gain, which allows you to change your frequency and hear the difference more easily. You can then identify the actual frequency or frequencies you want to remove.

Most professional EQ sections will have four different sections within them: high, high-mid, low-mid, and low. Each has control over a different range of frequencies, and each section has the three fundamental controls of a parametric EQ.

FILTERS

Filters are part of any decent EQ section (although they can sometimes also be located in the preamp section). They are very handy for cutting out unwanted noise and frequencies in a particular channel's signal. For example, you may want to add a high-pass filter on vocals to cut out any rumble and leakage that could be in the low end, or you might want to add a low-pass filter to your reverb to sit it back in the mix a little more without taking the level of it down.

There are a number of different types of filters. *Bell (peak)* filters are the most common type of filter; all EQs use them. A *high-pass* filter is a sweepable pot where you can select the frequency the filter starts (however, some consoles have a button that is at a set frequency, usually at either 80 or 100 Hz). *High pass* refers to the section that is not affected, so a high-pass filter doesn't affect any of the frequencies above the selected one. For example, if you select 400 Hz on this filter, everything under 400 Hz is cut from the EQ.

On some new types of digital consoles, you can select a slope for the cut. This is done as dB per octave, usually in the ranges of 6 dB, 12 dB, and 24 dB. (An octave starts at your fundamental frequency and ends when you have double the amount of cycles—so, for example, middle C on a piano is 440 Hz, the next C note above that is 880 Hz, and the C below is 220 Hz.) As you can see here, the higher the note, the more frequencies in the octave—so as you move up with your filter, you may want to increase the slope. As always, though, the most important thing is to listen to what is going on; this gives you the versatility to cut off frequencies with a steep slope or gradual slope.

These are hard (like 24 dB/oct) and soft (like 6 dB/oct) slopes. As a reference point, you may want to have a medium slope on a vocal, so you can keep the depth but also cut out some of the frequencies that make it muddy.

There are also *low-pass* filters, which work exactly the same as a high-pass filter, but in reverse; instead of cutting out lower frequencies, they cut out higher frequencies. This could be used to reduce hiss on things like a bass guitar, cut some of the extreme highs on vocals, and reduce sibilance.

Essentially, you should employ the low- and high-pass filters if you're using an instrument that will not produce frequencies (including harmonics) above or below the filter frequencies. This will help clear up any unwanted frequencies getting through.

Another type of common filter is a *shelf filter*. This will be on the low and high ranges of your parametric EQ. The frequency response of a shelf filter is shaped like a shelf or plateau. A shelf filter boosts or cuts a broad range of frequencies by an equal amount, either on the high side or low side of the audio spectrum.

The great thing about shelf filters is that you can use them in conjunction with your normal bell filters. For example, if there's too much bass in your bass guitar, popping your low EQ into shelf mode and reducing the low end by 3 dB will make the sound much smoother. However, there will still be too much 80 Hz in there, and you can't reduce the shelf filter anymore because you will then lose all the bottom end and drive in your mix. The solution is to use your low-mid bell control to reduce 80 Hz. Remember, gain is cumulative; thus, you won't need to reduce it by 6 dB, but only by 3 dB.

The bell (peak), high-pass, low-pass, and shelf filters are usually the only ones you'll find on a console. However, there are two more types you should know a little about:

- Band pass: A band-pass filter passes frequencies within a certain range. It's in the middle of low- and high-pass filters.
- Notch and band stop: Notch and band-stop filters are the opposite of band-pass filters. The center frequency is the center of the frequency band that is cut.

FREQUENCY RANGES

When talking about frequencies, we normally split them up into the following ranges:

- Lows: 20 Hz–200 Hz
- Mids: 200 Hz–3 kHz
- Highs: 3 kHz–20 kHz

As handy as these designations are, sometimes you need to be more specific. Here is a list of ranges and the label we give them.

- Sub: 20 Hz–60 Hz
- Bass: 60 Hz–160 Hz

- Low mid: 160 Hz–800 Hz
- High hid: 800 Hz–3 kHz
- Presence: 3 kHz–6 kHz
- Brilliance: 6 kHz–20 kHz

The "presence" and "brilliance" designations are fairly uncommon and are subjective descriptions of the perceived sound; "highs" refers to the frequencies that are affected. However, you will come across these words written on guitar amps, especially older ones, so you should be familiar with what they mean.

Inserts

When you insert an external device, such as a compressor or a gate, into your console, you need to make sure that the insert button is pressed. Some consoles will even let you choose where in the circuit you would like the inserted device to go. This means that you can manipulate the audio in different ways; for example, you could compress the audio, and then EQ it, or EQ it before compressing it. Note, though, that not all desks have an insert button; it depends on how the circuit works within the channel strip. Usually, the more expensive the console, the more likely you are to find this button. Insert sends can be used for multitrack recording, just like direct outs.

Send and Return

When you insert a device over a channel, the insert connectors are on the back of your console. Most of the time, there are two 1/4-inch jack sockets labeled "send" and "return" (though on cheaper consoles you may find only a single 1/4-inch jack socket labeled "insert"). These are known as tip/ring inserts and require an insert cable: a stereo jack to two mono jacks, which are labeled "tip" and "ring." The console will either send the signal on the tip or the ring of the stereo jack, and then return the audio on the one that isn't sending. The wiring configuration should be labeled on the desk.

Aux Section

The aux (auxiliary) section is used to send some signal from the channel strip to a monitor power amp or to an effects device (hopefully you should have at least four auxs). Some desks let you select whether you want to send pre- or post-fader on each aux send, and others have fixed auxs for pre- or post-sends.

If you are mixing FOH and are also mixing monitors from the same console, pop the auxs in pre-fader; otherwise, every move you make on the faders will affect the level in the monitors. This will likely upset the guys on stage—and potentially cause feedback—as you'll be constantly changing the levels in their wedges as you move the faders up and down.

Another thing to remember is to send your effects post-fader. That way, when you pull down the faders on your main mix, the level will also be reduced going to the effects, and you won't end up having a vocal swamped in reverb and no direct vocal.

Panning

The pan pot is normally located just above the fader (though with some desks, like the Yamaha PM4000, it is at the top). The pan pot control is your left and right control for sending to the left and right of the PA. It sends the signal from that channel to go to both the left and right speakers, or only the left or right speaker, or anywhere in between.

Routing and Bussing

Routing and *bussing* are the same thing. On each channel strip, there is a routing section, which is usually a grouping of buttons with numbers written on them.When one of these buttons is pressed, it sends the audio to what is known as a *buss*. The buss is a channel in the console that accepts several signals and combines them into a single signal. Similar to a bus that we take for public transportation, the audio gets on the buss at a certain point in the route, and then gets off at its destination. The destination, in this case, can either be your main mix (which is called your mix buss), your mono mix, any number of subgroups, or all of the above. (*Note:* Your mix and mono busses might be in a slightly different section as the rest of your routing buttons.) Routing several input channels to a group (subgroup) is useful because it means that you don't always have to send all your channels to the main left and right busses; you could, for instance, send all your drum channels to groups 1 and 2, compress them all together, and then send them to your left and right main busses.

When making your routing selections, make sure you aren't sending anything where you don't want it to go. For example, if you just want your audio to go to groups 1 and 2, make sure you don't have the stereo button pressed—this will send the audio to both groups. Sometimes, though, you may want exactly this; it is often purposely done, especially in small venues, in order to double-buss the vocals (to give you enough volume to get the vocal above the mix, which is the equivalent as tuning the fader up by 6 dB). It can also be used for *invisible compression (parallel compression)*. You mix an uncompressed signal with the same signal compressed and then reduce the level of the compressed channel. This enhances low-level detail without reducing peaks. The compressor adds to the combined gain only at low levels. This helps transients to keep their dynamics while still reducing the dynamic range. (I used this with Amy Winehouse because she has such a great dynamic range and was fairly controlled when she needed to be; in addition, the music lent itself to that type of sound.)

Mute

The mute button is usually one of the most obvious buttons on the channel strip; it's normally illuminated so that you can quickly see which channels are muted and which ones are not. Channels can normally be assigned to *mute groups*, which means that you won't have to individually unmute all your channels before the artist walks on stage.

Monitoring

There are two main types of monitoring on your console: PFL (pre-fade listen) and AFL (after-fade listen). This is the way you can listen to each individual channel over headphones without having to turn any other channels off. PFL and AFL do not affect the house signal.

PRE-FADE LISTEN (PFL)

The pre-fade listen (PFL) button is probably the most recognized button on the channel strip and is your monitor for any channel selected. It sends a signal from the channel you selected to your headphones or your listening wedge so that you can listen to that channel without having to listen to anything else. This button is also normally illuminated. Some consoles give you the ability to cancel all channels that are PFLed; this is in the master section and is usually accompanied by another, smaller button labeled "add" or "cancel last." This enables you to PFL multiple channels, or just one channel at a time.

On some consoles, the PFL button is called *solo*. In a studio console, the solo or PFL button affects the monitor-speaker signals. In live consoles, the solo button affects only the headphone signal. Otherwise, the audience would hear what you are soloing. You may also come across a *cue* button; this is the same as the PFL button and is mainly something you'll find on Yamaha consoles.

AFTER-FADE LISTEN (AFL)

The after-fade listen (AFL) button gives you the ability to monitor after the fader—so if you pull the fader down, you won't hear anything. You'll find these on the subgroups. This should be used for monitoring the signal sent to in-ear monitors (IEMs).

Solo in Place

The solo-in-place button doesn't appear much on most consoles, but is a very handy way to monitor individual channels through the PA speakers. As with actual solo buttons, where all other channels are muted, when this button is active, all the PFL or solo buttons become proper solo buttons, muting everything else apart from the selected channel. This is really handy during soundcheck, if you are having a few conflicts in the audio and want to hear what the individual sounds are like without having to mute all the other channels. (Note, however, that the band can sometimes get annoyed when the FOH PA cuts in and out when you press the solo-in-place button.)

Fader

Now we get to probably one of the most familiar parts of any mixing console; in fact, a mixing console doesn't look right unless it has faders. The fader controls the volume for each channel and is where most of the actual mixing of levels can be done.

Faders are a nice, easy, convenient way to feel your way around what you're doing; they are the connection between yourself and the music. I always have a finger on my main vocal fader, even when I'm not doing anything with it—it serves as a constant reminder of what I am listening to. I use this same method for all other channels as well. It's not that it actually makes it louder; it's just more physiological, but it is a really good tool to use.

MASTER SECTION

The master section is the most complex part of the console, mainly because the layout and labeling differ depending on the manufacturer. In this section, we discuss the most important information you need to get your show up and running.

Stereo

The stereo section is where you send your main left and right feeds to the PA. Also known as the stereo buss, this is where pretty much all your audio ends up.

Mono

The mono send is routed from your channel section and can be used to feed anything that needs to be in mono—such as a center cluster, or perhaps even your delays.

Subgroups

Groups or subgroups are mono or stereo. They sum several input signals and send them to the left/right master output buss. The subgroup as a whole can then be increased or decreased in volume by a subgroup fader. If you solo a subgroup over headphones, you can check the fader balances and panning of all inputs that are feeding the subgroup.

Most subgroups have an insert point where you can patch in equalizers or compressors and limiters that affect the subgroup mix as a whole.

Personally, I don't really like grouping, as I find that it adds noise to the signal path, but on digital console you don't have to worry about that. In some situations, however, it can be really handy. For example, you can assign many vocals to a buss and then use only one compressor inserted in that buss to compress all the vocal mics by the same amount.

If you have two kick drum channels, which is very common these days, you might want to compress them together. To do that, assign both mics to a group and then pan both your kick drum channels to one side—say, the left. This sends the audio only to the left-hand group, group 1. The pan on that group can then be put to the center; this way all the audio passing through that channel will go to the stereo buss instead of one side or the other. Then you can compress both channels with one compressor. The same applies to any other channels. Some consoles will let you select just one group, in which case not panning is needed, but others only let you select the groups as a stereo pair, as described above.

Voltage Control Amplifiers (VCAs) and Digital Control Amplifiers (DCAs)

On your mixing console, you may be lucky enough to have either voltage control amplifiers (VCAs) or digital control amplifiers (DCAs), never both. VCAs are on analog consoles, and DCAs are on digital consoles.

A **VCA** has no audio path for the inputs assigned to it. Instead, you assign a number of input channels to a VCA fader. That fader works like a remote control for all the assigned faders.

For example, you assign eight faders to a VCA group. If you turn down the VCA group fader, it's as if you turned down all the actual input faders. The balance between the faders stays the same.

VCAs allow you to control the volume of a group of channels without sending the audio through a summing circuit, thereby keeping the signal cleaner. You can also select multiple channels on multiple VCAs. For example, you can put all channels except for vocals on one fader, and then the fader next to that serves as the vocal VCA. This give you more control over the dynamics of the audio during the show.

Unlike with a group (subgroup), you can't insert a processor into a VCA group as a whole.

Matrix Section

On most professional consoles, you can send audio down matrix sends. These are similar to the auxiliaries but for your main outputs; you can also use them to feed delay speakers and mono clusters, as a stereo pair, or as a single mono feed.

Scenes

A scene is a "snapshot" or memory storage of the control settings in the console. Scenes are becoming more and more common on live consoles. They are an easy way of recalling specific actions during a show. For instance, when a new song is played, you are able to recall which effects are used, what setting is used on them, and which channels are muted for that song. This level of recall only happens on digital desks, although some analog consoles have a limited way of recalling certain parts like fader settings. They are particularly popular on theater consoles, purely because of the number of scene changes in the show and the need to turn mics on and off, but they are now becoming popular with rock and roll consoles as well. Scenes are usually found only on digital consoles, although there are some exceptions—the Midas H3000, for example. (*A little tip:* When using a Midas H3000, store the same scene above and below the one you are using. This way you won't do what I did at Bestival in the Isle of Wight: At the height of the song, my hand slipped off the fader and hit the scene recall button, which muted the whole thing. Luckily, no one on stage knew the FOH had cut out because it was so loud.

Talk-to-Stage and Shout

To conclude this section, let's briefly address communications to and from the stage. *Talk-to-stage*, or talk-back, is a mic line from the FOH console to the stage monitor speakers, which is a must-have accessory for the professional audio engineer. The line usually goes through the monitor console, if there is one.

The *shout* is a mic at both FOH and on-stage that is connected to speakers at either position and is always on, so the FOH engineer can talk to the musicians and vice versa.

OUTBOARD

Outboard is the term used to describe all the external processing equipment for your mixing console. We have given this topic its own section because it is easier to explain these processes separately from explaining mixing desks (and also out of love and respect for the old school methods).

Unfortunately, outboard equipment is becoming more and more obsolete as new types of digital processing equipment are developed; you can now have your entire FX rack contained in your laptop, so you need nothing more than your computer and a soundcard. Most digital consoles have all of this built into them—so with the push of a button, lights flash, the screen changes, faders fly around, and your graphic equalizer is presented to you. Another click of another button, and you have your reverb or delay unit staring you in the face.

Graphic Equalizers

When it comes to mixing, graphic equalizers are an audio engineer's best friend. Graphics (as they are more commonly called) are a very valuable tool with a lot of power. In the wrong hands, they can cause immense damage to a mix—but in the right hands, they can be an absolute lifesaver.

You can get different numbers of bands in a graphic EQ, but the majority are 31-band. This means that they have 31 separate volume controls, and each one is set to control its own frequency. The range is from 20 Hz to 20 kHz, which, as you should know by now, is the extreme range of human hearing. You will nearly always find that they have a switch that is labeled with $+/-6$ dB on one side and $+/-12$ dB on the other. This refers to the maximum-level difference between 0 dB and the top or bottom of each volume control. Most of the time it will be set to $+/-12$ dB.

Graphics can be a very personal thing when it comes to mixing. The more you EQ a PA, the more you will realize that there's a direct correlation between how you EQ the room and how you EQ your instruments. There will always be certain modes in a room that you have to take out, of course; just think about what you are listening to and what you are pulling out.

You should also be able to mix on the graphic. As previously mentioned, there are 31 different volume controls, each with its own frequency. I once worked a gig with the worst desk in the world; it was so slow, and I couldn't mix fast enough because of all the button pushing I had to do, so I began mixing on the graphic. As long as your gain structure on your console is near unity (meaning that all your inputs are coming out the outputs around the same level), this is a pretty easy thing to do.

Later on in the chapter on tuning we'll address the topic of using a graphic to EQ a room.

Dynamics

Dy-nam-ics {dahy-**nam**-iks}

—**noun**

1. (used with a singular verb) Physics. the branch of mechanics that deals with the motion and equilibrium of systems under the action of forces, usually from outside the system.
2. (used with a plural verb) the motivating or driving forces, physical or moral, in any field.
3. (used with a plural verb) the pattern or history of growth, change, and development in any field.
4. (used with a plural verb) variation and gradation in the volume of musical sound.

Dynamics are the range in volume of notes from loud to soft. Dynamics processors are devices or plugins that control the dynamics of a signal. Dynamics is slang for "dynamics processors."

Dynamic Processors

A dynamics processor is any piece of equipment or plugin that controls a signal's level changes (dynamics) or envelope. For example, if you have a bass guitar whose signal level varies dramatically, you might want to compress it to reduce its dynamic range; or if you have a kick drum that has a lot of sustain on it and you want to tighten it up, you might want to gate it, which would shorten its envelope. These are tools and should be used as such. A lot of people in the live industry don't really know how to use them properly, mainly because a live show isn't the time to play around with them. However, these tools can be extremely important to your mix.

The trick is to know when to use them and when not to use them. You need to have a deep understanding of what these things do to the music you are mixing and why they do it. You wouldn't compress a jazz band, but you might want to compress the brass section in a pop band. Why? Because in a jazz band, the mix is really controlled by the musicians, and their feeling comes out in how they play—your job here is more of a sound reinforcement one than a production one. For a pop band, however, you will want to make the brass section sound solid and together, like a wall of brass that you can easily place in the mix without too much interference with anything else that is going on.

When I was working with Seasick Steve, I didn't use anything other than compressors, and I only used two: an Avalon 737 (a very lovely preamp, EQ, and compresser) on his vocal, and a distressor (the best compressor ever as far as I'm concerned) on his guitar. The mic line from the stage would run directly into the Avalon. The output of the Avalon was fed to an insert in the desk. This way, the signal bypassed the desk EQ and preamp, so that you would just get the pure signal without any influence from the desk (apart from the fader). I used a distressor on the guitar as well and cranked up the signal up so that it clipped on the unit a bit. Basically, the unit was in a constant state of compression, and it made the guitar sound awesome.

Also with the Hoosiers, I took a Yamaha LS9 out on tour. I do like the ease and size of this console, but the audio quality is poor; it sounds like it's processing the 1s and 0s, and there is no warmth in it. As a compromise, I used a couple of Avalon 737s, running the console directly into the Avalons and then into the PA. I used the EQ section a little—it's a very lovely-sounding EQ—but not the compressor part. I was happy with the results; the whole audio just warmed up and came to life. (It still sounded processed, and the band plays very well and consistently, so it sounded more like a record than a live show—but that's what some people want.)

INSERTS

If you ever hear someone ask what inserts you want, they're referring to your dynamics processors. In other words, they want to know which dynamics processors you'd like to use in which channel. These are then inserted into the signal path. You can sometimes choose whether you would like your compressor before or after the channel strip EQ, but the inserts are always post-gain (after the input preamp but before the fader).

COMPRESSORS

A compressor is a device that controls signal level changes. If you have a bass player who doesn't play very consistently, loud and quiet notes can make it hard to get constant drive from your mix. By adding a compressor to your signal path, you can make the loud parts quieter (by compression), and the quiet parts louder (by makeup gain).

Compressors are fairly easy to understand once you experience them, but when you don't know what you're listening for, their effects can be a little hard to hear. Just keep playing with it, making extreme changes at first so they are obvious, then starting to make smaller and smaller changes. Listen to the difference.

LIMITERS

A limiter is a compressor with a compression ratio of 10:1 or higher. Limiters are roughly the same as compressors; the major difference is that compressors are meant to alter the dynamics but limiters are not. Limiters are designed to prevent signal levels from exceeding a threshold level that you set. This keeps signals at or below a fixed level. They're very common on the mix outputs for a studio mix, increasing the average loudness of the mix, although there can be some signal degradation. Also, most system management systems have limiters at the output stages of the crossover in order to protect the amps and speakers from uncontrolled high-level signals.

De-Essers

De-essers can be used to remove *sibilance*, which is the emphasis of "s" or "sh" sounds—hissing—that usually happens around 5–12 kHz. Some singers just have it in their voice, but other times it can be due to a mic's boosted high-frequency response. Either way, it's difficult to listen to, and it should be softened. I use a de-esser on the backing vocals for La Roux; when they were mixed down, they were tracked straight from the record, and that particular EQ doesn't always work every venue. In addition, the BVs are also EQed differently depending on the song. Using a de-esser softens the highs so that they sink back into the mix, rather than being thrown over the vocal when they come in. You will find that the controls are very similar to an EQ; de-essers have a frequency knob and a bandwidth (Q) knob, but they have a threshold knob as well.

DYNAMIC EQUALIZERS AND MULTIBAND COMPRESSORS

Unfortunately, in today's world, we have to work with budgets, so we need to be selective about what we take on tour. Everyone wants to take his own console, which is made much easier with the relatively cheap cost of digital desks (as compared to their analog forefathers). However, the ability to take your console requires adequate space in the tour vehicle. If you don't have your own console, you should consider taking a dynamic EQ. They are like the polish on the mix, so don't leave home without it.

A dynamic EQ and a multiband compressor are pretty much the same thing apart from one fundamental difference. A multiband compressor does not have the ability to add any expansion to the signal passing through it.

There are many benefits to using dynamic EQs, so it's always a great tool to have at your disposal. They enable you to de-ess the vocal while expanding the midrange, compressing the low end, and limiting the whole signal. They typically work over a 2- to 4-band range, although most have a button where you can defeat the filter (turn off the frequency bit) and use it over the whole signal.

The controls are essentially the same as the EQ section on your console, except that you also set your threshold and amount of compression or expansion.

You can also use dynamic EQs to select whether you want each frequency band to react above or below the threshold. This means you can EQ the signal as desired and then make it reduce when it reaches the threshold. You can even use it like a gate on a set of frequencies. This is extremely useful because there will always be certain frequencies that react differently in a room as volume goes up or down. This device allows you to smooth everything over and give you a nice polish to your mix.

As with all things audio, try it and see how it works for you.

GATES

A gate cuts out unwanted noise while an instrument is not being played, and thus can be a vital tool to create a clean mix. They're mainly used on drums, to avoid unwanted noise and rumble caused by vibration of the skins in between hits, but they are also used on each note's envelope as a creative tool. Some gates have a volume reduction control, where you can select how much the gate reduces in level.

EXPANDERS

An expander increases the dynamic range of the signal, as opposed to a compressor, which decreases the dyanamic range. Expanders make the quiet sounds quieter and the loud sounds louder. Expanders work below a set threshold—they operate only on low-level audio.

CONTROLS

Now that we've looked at types of dynamics processors, let's discuss their controls in more detail. All the information in this section can be applied to all audio equipment; these controls mean the same thing universally. The only thing that differs is what the unit is being used for. In Chapter 15, we'll look at how the threshold, ratio, attack, and release are used in relation to what we are mixing and how to set them effectively.

Threshold

A threshold is an activation point you set on a device; it is labeled in dB, just like a fader or meter. A gate, for example, will only allow a signal to pass through it when the signal level reaches over that set point. If the signal is coming into the console at −20 dB, and you have your threshold set to −10 dB, the gate won't open—but if the signal goes over −10 dB, the gate opens and lets the signal through. The same example can be used for a compressor, except that a compressor is activated at –10 dB.

Ratio

The *ratio* controls the amount of compression or expansion that is applied to a signal once the threshold has been crossed. The number is a ratio of the input signal coming into the device. A ratio of 1:1 means no compression, whereas a ratio of 3:1 means that the input signal must cross the threshold by 3 dB for the output level to be increased by 1 dB. Get your hands on the ratio knob and give it a turn to hear for yourself what is happening.

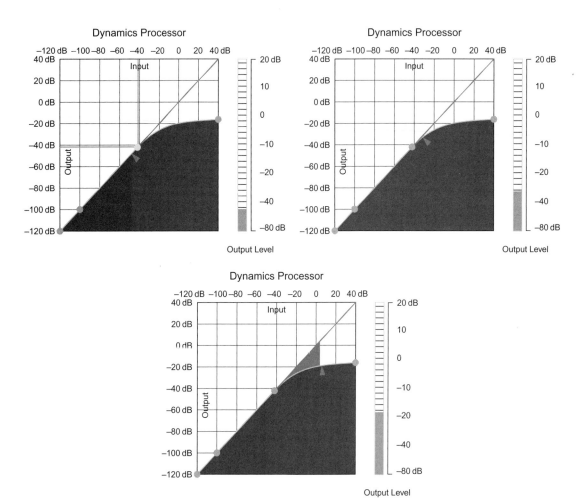

FIGURE 9.8
The yellow dot in the middle of the first screen is our point of threshold. As the arrow moves beyond the dot, the signal starts to compress. To see the color version of this image, visit the companion site at www.liveaudiobook.com.

1:1 is the lowest setting and means that nothing happens to the signal. ∞:1 (infinity:1) is the highest ratio setting and means that the signal level will not go above the threshold.

The compression ratio is "change in input level" vs. "change in output level" from the compressor. For example, suppose the ratio is 3:1. Then if the input signal level increases 6 dB, the output signal level increases only 2 dB.

Attack

This is how long it takes for a unit to react when the signal passes the threshold. Adjusting the attack time will help keep the transient information of the signal intact.

You can try playing around with the attack times on things like acoustic guitars. For a fingering, try adding more attack time, but if you feel you are hearing too much of the fingers you can reduce the attack time.

Release

The release knob controls how long the unit takes to stop reacting and can give you a lot of character in the signal. Short release times are good for short sharp sounds, like drums and percussion. Longer release times are good for things like vocals, where you don't want compression to be obvious.

Auto

The auto button is linked to the attack and release dials, and is usually only found on cheaper units. When this button is pressed, the unit automatically decides how to react to the signal. I would recommend that you don't use this; instead, set the attack and release manually.

Knee

The knee is the region between no compression and compression. A soft knee slowly increases the compression ratio as the signal level increases. With a hard knee, the signal goes into compression at a fixed threshold. You would use a soft knee on signals with higher ratios, which helps the transition between no compression and compression. See Figure 9.8, where after the point of threshold the orange line curves off to the right.

Gain Reduction

Gain reduction is the number of dBs that the gain is reduced by a compressor or gate. In a gate, if you don't want to completely turn off the signal, you can increase the minimum gain reduction so that the signal isn't cut off completely. This can be quite handy for things such as toms.

Peak and RMS Detection

Some units only allow you to use the attack and release knobs when the unit is set to *peak mode*. Peak and RMS (Root Mean Square) are two types of compression styles; they affect how the compressor detects how to compress the signal being passed though the unit. As you would expect, peak detection looks for peaks in the signal. For example, if you want to stop a snare from clipping the output, you could add a peak limiter to the snare channel. This will keep it nice and neat in the mix. RMS detection, on the other hand, works by looking at the average signal level going into the unit.

Stereo

Some dynamics units have a stereo or link switch, which links two compressors together, which can then be inserted over two channels that are running any stereo content. The advantage of this is that the compression in the two channels tracks each other, so that the stereo images do not shift during compression as they would if the two channels were compressed separately. Also, you only need to use the controls on the left-hand unit; threshold, ratio, attack, and so on, are all controlled on that side. The output stage of the unit is usually a separate control, but this does depend on the unit and the manufacturer.

Output Gain (Makeup Gain)

If you're having to do a lot of compression, you'll find that the signal is lower than it was without compression. Use the output gain knob to either increase or decrease the gain.

Side Chain

When you side chain a compressor, you're letting another signal control the unit. For example, if you have a compressor on your bass guitar, and then take the signal from your kick drum and feed it into the side chain of the compressor, the kick drum triggers the compressor. As a result, every time the kick drum is played, the bass is compressed; thus, you get a pumping sensation on the bass. This is used very rarely in live audio situations, but it can be used with more electronic and dance-based music. The one thing to watch out for is that if you are feeding a live kick drum into the side chain, and the kick drum is being played inconsistently, the compression on the bass will not sound good.

Transparency

When listening to something transparent, you won't be able to hear any type of compression. Transparency is directly related to peak and RMS detection. In RMS mode, your signal is more transparent and dynamic. Because the average signal is being detected, you have small varying-level changes, which give you more dynamics than in peak mode, but still keep the signal from getting too loud. In peak mode, peaks affect the whole signal and thus decide the overall compression of the signal.

Transparency can also be an issue with the quality of the components in the unit. Cheaper units use cheaper components, and this does affect quality. On these cheaper units, you'll find that when you compress a signal, you can hear it being squashed and forced through a hole. In this case, you might need to do some compensation after the compressor. I recommend Empirical Lab EL8x Distressors.

EFFECTS

Effects add the finishing touches to your mix, giving it more depth, width, and space. Once you understand what the controls do, you can produce all kinds of creative effects, from natural sounding reverbs to bizarre delayed flanger effects. There are endless possibilities, and, as with everything else, you should experiment so that you can hear for yourself what these do.

In this section, we'll look at some typical and atypical effects. We begin with the most common.

Reverb

Put simply, acoustic reverb is a series of echoes that come from multiple angles with close repeats, too closely spaced in time for the ear to resolve. It is made up of virtually an infinite number of reflected waves, all with different frequency spectra and delay times. We hear reverb nearly every time we hear a sound in a room, and it makes sounds seem natural.

The purpose of reverb units is to simulate natural sounding acoustic environments, such as a room in your house or a hallway. They can simulate unnatural sounding reverbs as well. Adjusting a few parameters on a reverb unit can give you a completely different sound. Each type of reverb has its own characteristics and tonal differences, adding much more than just a variety of reverb sounds to a mix.

Any reverb has two main parts:

- Early reflections: Sound reflections (echoes) that occur within about 1/20 msec of the original direct sound. They help to define the perceived size of the room that is creating the reverberation.
- Decay: The portion of reverberation after the early reflections when the echoes increase in number over time and gradually fade out or decay into silence. This is more of a "washy" sound; not much is defined. This then dies out as each sound reflection loses energy.

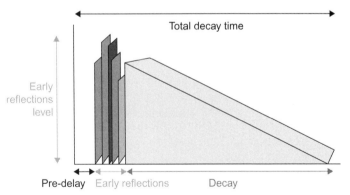

FIGURE 9.9
The initial reflections from the reverb are far more defined, but as the initial reflections themselves become reflected, the signal becomes very washy.

REVERB CONTROLS

Let's look at the main controls on a typical reverb unit.

Room Size

This is our virtual room. You can make the room bigger or smaller, depending on the kind of sound you are looking for. This puts a set of variables into the reverb unit, which affects how the soundwaves travel around the room. Be careful not to set the unit for too small of a room, as this can cause a warbling or a fluttering effect, which can sound very unnatural.

Pre-Delay

This is the control for the time it takes for the first repeat to come back. When you're in a bedroom, for example, the natural pre-delay time would probably be somewhere around 5–10 ms—in other words, it's nearly instantaneous. This is very different from being in a large space, such as an arena; the pre-delay here could be somewhere around 40–50 ms, or maybe even more. Pre-delay determines the perceived size of the simulated room. With pre-delay, you will hear the initial (direct) sound first, followed by the reflected (reverb) sound shortly after it. This gives us our sense of space in any given environment. Using pre-delay on a vocal can clarify the vocal by separating the direct sound from the onset of reverberation.

Reverb Time

The reverb time is also known as the *decay time* on some units and is the time it takes for the reverb to die off to 60 dB below the original sound level. In a bedroom, you probably wouldn't notice any reverb at all because the reverb time is so short—probably around 0.2–0.3 second, maybe less. In an arena, however, you could have a reverb time of 5–6 seconds, or even more. Obviously, these are two extremes, and there are many variables in between the two; you can have a reverb time from 0.1 second right up to 99 seconds.

Diffusion

Diffusion is the space between the reverberated repeats, and it tells you about the kind of space you are in. Being in an average room will cause low diffusion; this is because there are likely lots of soft furnishings, which means less sound reflection, and thus a longer time between repeats. High diffusion would happen in, say, a cathedral, where there are stone walls and hard surfaces and lots of surface irregularities to diffuse (spread out) the sound reflections.

Density

The density controls the very first short delays of the reverb. As we saw earlier, the early reflections are more defined than the later ones. This setting controls the amount of early reflections, so you can create clarity in the vocals, or blend and soften the hit of a drum. Here is a general rule of thumb: more percussive sounds will require higher densities, and more melodic sounds will require lower densities.

Filters

Many reverb units have settings to change the EQ of the reverb. These aren't complex EQs, however; they're normally quite basic, with three bands: low, mid, and high. Each usually has a frequency control, a bandwidth, and a gain. The bandwidth on the high and low can usually be turned into a shelf or pass filter. This kind of EQing gives the reverbs their tonal characteristics.

TYPES OF REVERB

There are many types of reverb presets to choose from, such as "Vocal Verb 4" or "Extended Church 4.5." Below we explain a little about each of these types.

There are two important things to remember as you read through this information: (1) Not all types of reverbs suit all types of room; and (2) you don't always need to add reverb. In other words, let the natural reverb of the room work for you. Adding extra reverb only where and when it's needed will result in a crisper and more well-defined sound. Many venues have lots of natural reverb, so you might not need to add any reverb to your mix—it will just muddy the sound.

Vocal

Vocal reverbs are fairly rich in sound, and not too long—maybe around 1–1.5 seconds of reverb time, and a shorter pre-delay of around 10–15 milliseconds. They have a lower density than most other reverbs.

Note that just because this is called a *vocal* reverb doesn't mean it has to be used on a vocal; in fact, I frequently use it on drums. Remember: When something sounds right, it is right—no matter how you got there.

Room

Room reverbs are fairly short reverbs without a lot of pre-delay. Most units have a number of different types of EQ on these types of reverbs: phone booth, bathroom, kitchen, and so on. Rooms with lots of soft surfaces don't have many high frequencies in them, whereas rooms with hard and shiny surfaces reflect more high frequencies.

Hall

As you can imagine, hall reverbs are bigger than room reverbs and are usually the smoothest type. They have longer decay times and a larger room size setting, creating more complex reflections. This means that when the reflections blend together, the decay is a lot smoother.

Plate

A plate reverb is the brightest sounding of all the reverbs. It's used quite a lot with vocals, though, in my personal opinion, I don't like having that much brightness in a reverb on a vocal. I find that it tends to distract from the main source sound. (You could always turn the top down or add a filter in this case, however.)

A plate reverb is similar to a spring coil reverb in a guitar amp, which is created by a spring mounted in a metal box with two transducers at either end. (A transducer is a device that converts energy from one form into another, such as a guitar pickup.) The signal is fed into one side and passed through the spring, which starts to vibrate in the metal box. When the signal reaches the other end of the spring, it gets converted back into audio, but energy from the signal stays in the spring and keeps on vibrating until it dies away—this causes the reverb sound. The plate reverb uses the same principle, except with a metal plate rather than a spring, and the transducers are placed on the plate, rather than at either end of the spring, and these days they are emulated in a digital unit.

Gated

A gated reverb is literally a reverb with a gate on it. Gated reverbs have an intense reverb for a short time, and, rather than dying off slowly, the gate comes in and just turns the reverb off. This was very popular for snare drums back in the 1980s, so it's a technique that can sound a little dated if used. But in the right context (as with everything), it can sound great. I personally quite like gated reverbs, especially if you have a second snare that is much higher pitched than the main snare. They can also sound great on vocals and guitars, but instead of using a gated reverb program, you can create your own verb, put a gate over the verb returns, and then side chain the instrument into the gate. This way the gate is opened only when the guitar or vocal is being used.

Delay

The other commonly used effect is the delay unit, which creates discrete echoes. It is sometimes marked with the abbreviation DDL (which stands for digital delay line), but most of the time the name of the unit is written on the console.

Delays are fairly easy to understand: you feed a signal into the delay unit, it holds on to the signal for a little while, and then it sends it back. Unlike reverbs, where you are getting multiple repeats that increase in density over time, you just get the same signal repeated over and over again until it eventually dies away.

The sound that is produced is an echo. A delay or a repeat is not a sound, but they are both widely used terms when referring to the actual sound that comes from a delay unit.

DELAY TIME

Delay time is the time interval between the direct sound and its first repetition from the delay device. It is also the time interval between multiple echoes. You can set how long you would like the delay to be, and it is measured either in beats per minute (BPM) or milliseconds.

Most delay units have some form of tap button on them, enabling you to tap to the music and get a delay time that matches the tempo of the song. The trick here is to make a delay sound a little more interesting by speeding it up and slowing it down for different parts of the song, changing the repeats. Remember, though: Just because you have it at your disposal doesn't mean you need to use it. Think about the song, the artist, and their style of music before going crazy with it. And if you have a band that is working to a click track, then you might not want to put the delay in the exact time; otherwise the music might feel very regimented and not flow naturally.

FEEDBACK

The feedback setting controls the number of repeats. It works the same as a feedback loop in an audio circuit, but instead of sending squeaks or rumbles into the room, it takes the output of the delay and puts it back into itself, giving you more and more repeats as you require it. When no feedback is applied to the delayed signal, you only get one repeat.

NUMBER OF REPEATS

Some delay units have a *number of repeats* function. Like feedback, this gives you control over the number of repeats as they fade to silence. This give you more control over the repeats than feedback does and in most cases overwrites the feedback function.

You can have two repeats with a feedback of 95%, but your delay unit will still only repeat twice.

FILTERS

Just as with reverb units, some delay units have filters. This function can be really handy when you want to simulate a tape slapback echo, which has reduced highs with each repetition.

SLAPBACK DELAY

Slapback delay is very common. It's very similar to the echo, but the repeats are slightly longer (about 120 msec between repeats), so the sound is slightly more defined. Slap echo was first used on 1950s rock and rockabilly songs.

Modulation Effects and LFO

Modulation effects are any effects that change the direct sound. Unlike reverb or delay, where the direct sound stays the same and you are just adding to it, modulation effects use delay to alter the direct sound's pitch. Some modulation effects use a device called an LFO (low-frequency oscillator), which produces frequencies that are so low that they can't be heard (and thus can really only be used for modulation effects). An LFO typically applies a slow variable delay, changing the pitch and causing chorusing when the direct sound is combined with its delayed replica.

Modulation effects are used quite a lot within studio mixes, but a lot less within live shows. If not used correctly, they can cloud the mix—but when used correctly on the right type of instrument, they can bring warmth, color, and variation to your mix. Next we'll have a look at some types of modulation effects.

PITCH SHIFT

Let's look at the pitch shifter first, as it is probably the easiest to explain; it is used for thickening up weak vocals or for adding some extra uniqueness to your vocal. An example is the fine-tune setting, which is the small frequency differences between each note rather than moving your pitch a whole note up or down: I just want to make it sound a little thicker, and where the notes are so close together they start to interact between themselves and create a warbling effect. With La Roux, I use +4 on the left side and a delay of 14 ms; on the right side I use −14 and a delay of 4 ms. When combined with the vocal, it gives it a very unique sound and sinks the vocal into the music while still keeping it loud enough to be heard.

CHORUS

A chorus effect usually has a variable delay time of between 15 and 40 ms. The delay time causes pitch variations in the incoming audio and thus is supposed to sound like a choir. You can vary the intensity, which is sometimes called *modulation*, and also the rate, which is the frequency or speed of the pitch changes.

Because of the nature of the chorus effect in creating a simulated choir of instruments, it works really well on vocals, guitars, and keyboards. The effect can add a richness and fullness, giving movement to the audio. It sounds good on long sustained notes.

Flanger

In a flanger, the direct sound is combined with itself delayed less than 20 msec, which creates a comb filter. The delay is continuously varied by a rate set by the LFO. The delay varies between 0 ms and 20 ms. Unlike the chorus effect, the signal can be fed back on itself. The variations in the delay cause the comb-filter notches to move up and down in frequency, which gives a sweeping effect and adds some really interesting color and variation to your signal. It produces somewhat of an underwater, spacey effect.

I frequently use this effect on vocals for rock bands that have longer vocal notes; it thickens up the vocals a little more, and it gives boring vocal lines some life. It works well on guitars and keyboards as well.

Computer-Based Effects

Computer-based processing systems can bring much more versatility to your live show. In addition, because they are so small and light, you can take them anywhere—no more lugging a big heavy rack through a muddy field to get to the FOH compound of a festival. Another massive advantage of computer-based processing systems is that you can load the exact same settings a band used to record their album, so you are now even closer to re-creating the album live.

My own computer-based system consists of a midi-controller with faders, a sound-card, and my Mac. I run a program called MainStage within which I keep all the effects that are needed for a show. The midi-controller controls all the virtual faders, and also has midi-program changes on quick buttons on the top of the unit, so I can change to a different song's effects at the tap of a button. Because the controller is portable, it can sit next to me on the console for easy access at all times.

The only major disadvantage of using this type of system is the latency caused by the computer. Reducing the buffer size within your computer can help, but in dead rooms (rooms with next to no acoustical reflections), you can still hear a slight delay.

Playing with Effects

It's a great idea to play with effects as a way to understand what they do and how they respond to various different input signals. For example, feeding a delay back into itself can give you a kind of dub delay, where the signal starts to decay into noise. (Do be careful with this, though: Sending too much signal back into itself will cause your head to explode, along with various channels of smoke being vented out of the device.)

Remember: You aren't restricted to standard effects. There are many combinations that you can use, and they can be applied in various ways. You may even discover some brand-new ones of your own.

CHAPTER 10
Line Systems

After you've put up your desk and outboard racks, you'll have to run out your multicore (or at least get one of the very helpful local crew to run it out for you). Multicores are also known as *snakes*, or sometime just referred to as *the multi;* they are the method of getting mic lines from the stage to each console, and then getting the audio back to the on-stage power amps and monitor console. (We'll go into more detail about multicores later in this chapter.)

In professional systems, the signal paths work as follows:

- Input Source: Vocals, guitars, keyboards, etc.
- Microphone or DI box: we'll go into these in the next chapter when we talk about setting up the stage.
- Stage box
- Splitter rack
- Multicore
- Consoles
- Returns

To properly understand how and why these systems work, and why we use them, we need to first look at *unbalanced* and *balanced* lines.

UNBALANCED LINES

Unbalanced line cables are frequently used for connecting guitars and keyboards into a PA system. Inside, they have a single insulated core, around which is wrapped a screen, which is a mesh of wire. This screen acts as a shield against electrostatic hum fields and RFI (radio frequency interference).

These kinds of cables normally have a mono quarter inch jack socket on them. You can identify the difference between a mono and a stereo jack by the amount of rings it has on it; a mono jack has a single tip that separates the single core from the screen, and a stereo jack has a tip and a ring. A mono jack system is great for connecting things over short distances. However, for cable runs over about 18 feet, the cable starts to act like a massive aerial and will pick up more noise than you want in your circuit. This, of course, is a big problem if you're trying to send audio to the front of house console and then back again—they are usually more than 18 feet away.

To solve this problem, you must use a *balanced line*.

BALANCED LINES

The majority of professional equipment, including microphones, mixing consoles, and outboard, use the balanced line system. But what does it do, and why do we use it? And, most importantly, how does this affect our mix?

To explain this concept, we need to first look at what *balancing* is. The basic idea behind a balanced line is that, unlike its unbalanced counterpart, it rejects any hum fields that can be picked up by the cable. To understand why, think about the cable that carries the signal. The cable consists of a pair of wires twisted together, known as a *dual core*. The pair have two different colors; one is referred to as *hot*, and the other as *cold*. They are wrapped together so that they are both subject to the same *interference* (electrostatic noise picked up by the cable). They are then wrapped in a screen, which is a mesh of wire that surrounds both internal cables and shields the cable carrying the audio from external electrostatic fields.

Each of the two wires inside the cable carries the same waveform information, except that one of the waveforms is exactly 180 degrees out of phase (opposite polarity) with the other. (Recall from Chapter 2 that, if two identical signals are opposite polarity and summed together, the signals cancel out.) Microphones and mixers have a balancing circuit in them. This circuit is used to convert the signal on one core coming from the microphone to the reverse waveform, which is the opposite polarity of the signal in the other core. Then, when the signal reaches the mixing console, the mic preamp converts the signal so that both signals are back in phase, after which you end up with a signal in which any hum picked up is canceled out.

Along with the two wires that carry the signal, we also have a screen just like the one we looked at in the unbalanced line. To make a balanced cable work properly, you need to be able to connect the three wires to three connector pins. Most professional equipment uses XLR-type plugs and sockets; this is a three-pin connector where the second and third pins are your hot and cold wires, and the third pin is your screen. Occasionally, you'll come across stereo quarter-inch jack plugs and sockets. As discussed before, a stereo jack can be identified by the two rings (the tip and ring) it has on it, which separate the cores from each other, and also the screen.

Using this system has other advantages in addition to canceling noise. Because you have two sets of the same waveform, the two cables combine their signals at the mic preamp input. This means doubling the amplitude and thus getting a stronger signal. This is why, when one of the hot or cold wires is cut, we lose 6 dB in signal (half the amplitude).

The advantage of all of this is that we can send a good, strong, clean signal down even a long balanced line cable without picking up hum; the high-frequency response of the signal is only limited by the capacitance of the cable. The distance along which you can get a clear signal is usually about 0.2 mile, which is almost certainly longer than the distance to the stage you'll be mixing on.

MIC AND LINE LEVEL

In the professional audio workplace, you will come across many different types of signal strengths—for example, the signal produced by a guitar or a power amplifier. The same goes with microphones and other on-stage instruments, such as keyboards or soundcards. The signal strength from microphones is called the *mic level*, and the signal strength from instruments such as keyboards and soundcards is called the *line level*. Mic-level signals are just around 2 millivolts (up to about 1 volt for very loud sounds with sensitive mics). Line levels, however, are the levels of signals that have been amplified and measure around 1.23 volts; clearly, there is a massive difference.

Because on-stage signals are going into a mic input, all unbalanced, high-impedance line-level signals from instruments are converted into balanced, low-impedance mic-level signals (by using a DI box) and sent from the stage down the multicore. (Sending a line-level signal into a mic input might cause it to overload.) Once the mic level has been received at the console end of the multicore, mic-level signals need to be amplified into line-level signals, which is the level sent to the power amplifiers. Any piece of equipment with a pre-amp is working with mic-level signals to amplify them up to line-level signals, which are processed and routed in the device.

To increase the level so that our mic level can work with our line-level electronics (mixing console, amplifiers), you need to put it into a preamp. Because mic-level signals are quite weak, it's important to have extremely good pre-amps. Keeping the quality of your signal is essential at the front end of the signal path because any distortion in the preamp will only be transferred and exacerbated farther and farther down the line.

STAGE BOXES AND SATELLITE BOXES

When you want to plug in multiple mic lines, it's best to keep everything as neat as possible; the best way to do this is via a *stage box*. A stage box is any on-stage box with XLR connectors for your microphones. From those connectors, the mic signals travel down a signal multicore cable. A *satellite box* or sub-snake is a similar concept; it is a smaller stage box run to different parts of the stage. It can run all the lines into the front of a drum kit or keyboard, instead of all the way across the stage. You can run your smaller satellite boxes to one main stage box, patch your XLRs into there, and send their signals to the front of house.

SPLITTER BOX

If you get to the very fortunate position of not having to control monitors from the front of house, you will need to get a separate monitor console. This means that you will need to split all your lines from the stage to a monitor console and the front of house console. In other words, you need a splitter box. In most cases, the splitter box and the main stage box are the same piece of equipment, and all your lines from the stage are run to this point. If you are only splitting the signal once, giving you two lines, you can use passive splits. However, if you have to split again (for a recording console, perhaps), you should use active splits. The difference between passive and active splits is that passive splits don't require power, whereas active splits do. Using active splits helps reduce signal degradation on the lines.

MULTICORES

A multicore is a bundled group of signal cables contained in a common jacket. Each cable inside the multicore carries a different signal. The multicore carries your mic signals from one place to another, usually from the stage to your consoles at the front of house. Just as balanced line cables have a single twisted dual-core pair inside, multicores have multiple twisted pairs inside them. They also have big multipin connectors on each end of the cable, which then break out into tails with XLR connectors (also called *breakouts or fan-outs*). These tails are the multiple lines within your multicore that plug into your console, such as mic cables.

Multicores come in two types, analog and digital. Analog multicores and their associated equipment are quite heavy; they carry the analog mic signals down the cable. Digital multicores, however, only carry data and are much easier to work with. Because they simply transmit data instead of sending audio down them, they are nowhere near as heavy as analog multicores, take very little time to plug in, and use up very little space. Also, they pick up much less interference than analog multicores. They usually use a Cat5/6 Ethernet cable to transmit their data. Although there are a couple of different types of digital multis, they are all relatively similar in format.

If you do end up working with analog multicores, you should be familiar with the *returns multicore*. These return any audio from the desk to the stage—like the main mix or monitors sends to be sent to the power amps on stag, and any matrix sends. Occasionally you might come across the system crossover at the front of house; if used on an analog multicore, the sends will be split into the different frequency bands and sent to the power amps on the returns multicore.

Power is another thing that will usually be run next to your front of house multicore. Be sure to keep any power lines at least 1 foot from the multicore so that the mic signals in the multicore don't pick up hum radiated from the power lines. It is important, as we saw before, that the PA system and anything connected to it are all on the same power circuit. Because of this requirement, you run the power from the main power distribution unit on stage.

CHAPTER 11
Acoustics

Everything we listen to has acoustical elements to it. Recall our discussion of reverb: Reverb is a simulated acoustic environment created to make your mix sound more spacious, and it can give the effect of being in a certain type of venue (for example, a church or a canyon). Reverb is in everything we hear, and sounds would seem very alien without it. If you listen to an in-ear mix, you can get an idea of what it's like. It can be hard to work with, because we're used to hearing sounds interacting with their surrounding environment.

The acoustics of live environments are hard to predict and very complicated. You may find yourself in a venue where the sound is bouncing off all the hard surfaces in the room, and there is nothing you can really do about it. In other cases, you may have finished your soundcheck with everything sounding fine, and then find that the room tightens or loosens after the audience enters. Although the topic of acoustics could be the subject of an entire book, in this chapter we focus on the information you need to command a successful show. It is important for you to understand why in some places you struggle to get your mix together and in other places you don't. This information is relevant to all types of rooms, audiences, and PAs.

UNDERSTAND THE SPACE

The first thing you notice when you walk into a venue is that it's a big empty space covered in either concrete, brick, wood, glass, or a combination of all of these. Some places may have hard, shiny surfaces (which is common in the modern-day architecture of a musical performance area, but can be difficult to work with). If you're lucky, though, the venue has been well designed, and the acoustics will be perfectly manageable. For example, Amsterdam's Heineken Music Hall is essentially a large square box, and, theoretically, should create standing waves and phasing problems. However, it doesn't, because the designers integrated a wonderful system of acoustic dampening.

When we say a room is *live or lively*, we mean that its acoustical properties are quite active; in other words, there are shiny, hard surfaces and flat walls, both of which cause a lot of reflection. When a room is too lively, it's hard to get definition; hearing what you're doing becomes tiresome because you must concentrate on what you're listening to.

At the other end of the spectrum are dead rooms, which are rooms with next to no reverb in them. This might be due to acoustical treatment, or simply soft furnishings. In these types of rooms, you can feel a kind of pressure in your ears; because you're used to hearing reflections (even when unaware of it), sound seems quite unnatural without them. When mixing in a dead room, it's hard to make the sound seem natural because all the elements that would normally do this are missing. Not even recording studios are completely dead because you want to be able to capture some kind of naturalness. In a dead room, you can hear every little thing you do. This can be good, but it can also emphasize all the flaws in your mixes. You will likely need to add a fair amount of reverb into your mix to get a natural feeling.

No room is perfect—it's really all about working with the room rather than trying to make the room work for you. A room doesn't change its sound to fit a show; it's the speakers that change their sound to fit a room. The room responds to what you put into it, and what you get out of the room is a direct result of your actions—if you push too hard, it'll just push right back. If a venue is very reverberant and muddy sounding, it can help to use highly directional speakers aimed at the audience so that the sound is absorbed by the audience and not reflected so much by the ceiling and walls. Of course, some rooms are easier to work with than others; these are generally those rooms that have few reflections and reflections that are absorbed fairly quickly. Rooms should also have enough space for your mix to breathe, rather than it feeling oppressive and like it's coming at you, which it can do in some dead rooms.

WAVELENGTH AND STANDING WAVES

Recall the topic of pressure zones when we looked at where to position mixing consoles. These are areas where sounds nearly double in energy (the only reason they don't totally double is because they lose energy as they hit the wall). A 1,000-Hz soundwave has a wavelength of approximately 1 foot. With every doubling of frequency, the wavelength (the physical distance between sound compressions or peaks) is halved—so the wavelength of a 2,000 Hz wave is half of a foot, and so forth. However, when you halve the frequency, the wavelength *doubles*. So 500 Hz has about a 2-foot wavelength, and 250 Hz has about a 4-foot wavelength.

When a sound system produces frequencies below about 300 Hz, a lot more energy is involved in moving the soundwaves; it takes bigger amps in your amp rack to make soundwaves of these sizes at a high level. Because these frequencies have long wavelengths, any kind of minimal acoustic treatment has very little effect on them because the treatment has to be at least 1/4 wavelength thick to be effective. When we start getting to frequencies around 20 Hz, we are looking at wavelengths of around 56 feet—that is, you are playing with dimensions that are actually the sizes of rooms.

The best way to understand a standing wave is to think about it visually. If you could actually see soundwaves, and you were to look at a 20-Hz cycle, you would see a waveform 56 feet long. If you had a room 56 feet long, you would see the entire cycle fit nice and neatly within the two walls of the room. However, when the soundwave hits the wall, it gets reflected back. If the sound-wave's reflection follows the exact same path that the original wave took, you wouldn't see any physical movement at all: hence the name *standing wave* (which can also be called a *room mode*). If the wave is reflected back in phase at a certain point in the room, you hear an increase in volume of that frequency; if it is reflected back out of phase at a certain point in the room, you don't hear that frequency at all. So standing wave can make certain bass notes much louder, or much quieter than others.

Standing waves can occur in either full or half waveforms, as long as they fit exactly within the walls of the room. If your room is 28 feet long, you can still get a standing wave of 20 Hz; similarly, if your room is 112 feet long, you can still get a standing wave of 20 Hz. Also, several frequencies create their own standing waves, so a room can have dozens of different-frequency standing waves at the same time.

DISRUPTING SOUNDWAVES

In general, soundwaves are quite chaotic, which is how we hear them. When you start to get waves that add or subtract, that's when you run into problems. Parallel walls, for example, can build up acoustical energy, nearly doubling frequencies in volume (they also cause flutter echoes that bounce back and forth between the two parallel walls). Concave structures, on the other hand, focus all the acoustic energy in one point. Both types of surfaces can be difficult to work with.

Controlling reflections is far better than trying to eliminate them. Listening to anything that doesn't sound quite natural can cause listening fatigue, so you need to disrupt the way the soundwaves travel through the room—by pointing them in directions where they need to be (toward the audience) and absorbing them (or using controlled dispersion) where they don't need to be (toward the walls and ceiling). Multiple arrivals of the same waveform cause a clouding of that soundwave, making it less clear than it should be. The trick is to not to have lots of reflections of the same source sound. The only way not to create standing waves is to build carefully tuned bass traps into the venue walls or ceiling, and of course you can't do that as a traveling sound engineer. Fortunately, standing waves are much less of a problem in large concert halls than they are in small rooms. Repositioning the subs can affect the strength of the standing waves that they create.

LOW END

Bass is one of the elements we listen to most when we are at a live show; when it is missing, people often have trouble getting into the music. Even when listening to a classical concert, low end is involved; most of the rhythm parts are played on lower register instruments. The bass guitar is also considered to be a rhythm instrument. When bass is controlled well—incorporating all the frequencies that need to be reproduced, from those that give us punch to those that we can only feel—it can be extremely pleasant and can even draw you closer to the music you are listening to.

That being said, bass frequencies are some of the hardest frequencies to control in a live environment, and it is due to bigger soundwaves that contain more energy. Unlike high- and midrange frequencies, which are more easily absorbed, getting smooth bass can be very difficult. If you're having problems with this and have enough space to do so, try restacking the subs in the system. A sub itself is fairly omnidirectional, so moving the subs around isn't going to affect your left and right mix. What you are trying to achieve with this is good coverage of bass across the venue, without creating a point where the bass sums together.

Different rooms react differently to bass frequencies, so you need varied techniques to get the best out of what you're given. Normally, you'll find that subs are stacked left and right, underneath the high and the mids. With this type of configuration, you're more likely to get a summing of low-end frequencies because of the coupling of these frequencies in the middle of the room. As long as your system has been set up correctly, all the frequencies should be in phase—this is where you're more likely to start seeing standing waves form. If they do form, the ideal solution here is to move your subs, which moves the point at which the summing occurs. I think that subs placed in the center of the room work the best in most environments, as long as space is available. This creates fewer modes within the room and helps produce an even sub sound across the entire floor.

Another possible configuration of subs is an arch, which can either be in the center, left, or right of the room. The back of the subs will be closer together, and the front will be wider, causing the sub to flare out into the room. This technique can take a while to set up, as you need to make sure that the spread is equal on both sides and that the space is equal between the subs. However, this can cause phasing issues, and the benefits of getting the even spread of the low frequencies don't always outweigh the drawbacks.

Another possible solution is to place the subs, spaced equally apart, across the front of the stage. This method works really well, but you do need to make sure that the subs are all exactly in line with each other and that they really are equally spaced; otherwise you will just create smaller pockets of bottom-end summing.

Many subs use a horn-loaded system, where the bottom-end is pushed out to the left and right. By turning them on their side, you can focus the majority of the push down to the left and right of the room, so the radius of the low-end frequencies is narrower. This way, if summing occurs, it should be dramatically reduced. This technique is great if you're dealing with a balcony; it throws low-end right to the top of it, but it can also cause problems depending on the size of the room. If you have a long balcony, you might find that the low end rolls back on itself underneath the balcony, causing phase cancellation. Also, throwing that much bass into an open space can excite the acoustics of the room.

MATERIALS
Reflections and Absorption

Pretty much everything reflects soundwaves; the point is how *much* sound is reflected from the material. Different materials reflect different frequencies. Drywall, for instance, absorbs higher and lower frequencies better than mid-range frequencies, so when you point a full-range signal at drywall, you will get more mid back than other range of frequencies.

Take the Rough with the Smooth

Another thing to bear in mind is that smooth surfaces reflect more sounds than rough surfaces. Smooth surfaces have all their surface area pointing in one direction, making it easy to reflect sound; however, rough surfaces catch sound and reflect it in different directions, An unvarnished wood floor, for example, reflects sound, but does so even more when it's varnished. Cardiff University's Great Hall had acoustic panels installed to cut down the reflections that were caused by a huge wall opposite the stage, directly in line with the PA. However, they were painted white, which defeated their purpose. If something looks shiny, it is probably smooth and will reflect more.

Bending Soundwaves

The bending of soundwaves is known as *diffraction*. Soundwaves bend around objects, depending on the size of the wavelength. Higher frequencies have a shorter wavelength; thus, when a frequency's wavelength is shorter than the obstacle it comes into contact with, it is absorbed or reflected by that obstacle. Lower frequencies have a longer wavelength, which bend around the obstacle and rejoin on the other side. To understand this visually, imagine throwing a stone into water: When part of the resulting ripples come into contact with a rock, that part of the ripple is stopped, but the rest of it continues around the rock and rejoins a small distance away on the other side.

Transmission

Moving molecules in the soundwave contain energy. When these molecules come into contact with an object, be it a wall, window, or person, part of that energy is transmitted through the object. Massive walls transfer less sound energy than lightweight walls. That's why a cement wall blocks sound better than a piece of drywall. Low frequencies are transmitted better than high frequencies. This is why you can feel and hear the bass from the next room, but not hear the highs.

Reflection of Different Materials

Naturally, many different types of materials are used to make up the interiors of venues. In this section, we'll look at a few of the reflections you're most likely to encounter.

As a general rule, low-end frequencies are reflected by most things because lows are difficult to absorb. The wavelengths are so large that, due to either reflection or diffraction, they are difficult to eliminate. When you want to absorb as much lower-end frequencies as possible, it's best to use a material that vibrates easily (like a thin wooden panel); this absorbs more low-frequency energy than, say, a solid wall. For example, materials such as drapes offer the best type of absorption. Carpets absorb only high frequencies. Drapes are better than carpets because they tend to be hung, and the vibrating drape takes some of the energy out of the frequencies. Also, a drape with pleats or folds has more surface area to absorb sound than a flat drape does. If a drape (or other absorber) is hung 1/4 wavelength from the wall, it absorbs much of the frequency having that wavelength.

Interestingly, windows are good at absorbing low frequencies because of the movement they inherently have in them, but they are also smooth, flat surfaces—so mid- and high-end frequencies bounce off them like a rubber ball on a brick walls, and thus they aren't a viable acoustic treatment. In some venues, you may see glass barriers surrounding balconies; these tend to be toughened glass. The harder the material, the harder it is for lower frequencies to be absorbed, so they reflect more lows than more flexible glass does.

Other materials, such as wood, reflect more midrange frequencies, while some of the higher-end frequencies are absorbed. This can result in a high-mid honk (but when the floor is full, that reflection is reduced).

Bricks tend to reflect more than they absorb. They can absorb a little more high-end than lower and midrange, but not enough to make a massive difference. Brick is fashionable at the moment, so many venues now have exposed brick. Unfortunately, though, this is never a good acoustical choice.

The acoustic properties of concrete depend on how it's finished. If left to set in its natural state, it will be like brick and will hold some of the high-end frequencies back. Most of the time, though, it's smoothed down and varnished.

Now that you understand the basics of material acoustics, let's look at the more practical elements of what you can do with the acoustics in the environment you are given.

VENUE ACOUSTICS

As part of a touring show, this is something that you're going to have very little control over, but if you're working in-house, it's something you should seriously consider. Whatever environment you're in, you should understand how the sound translates throughout the room and how you can get intelligibility

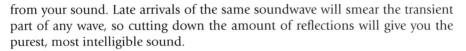

from your sound. Late arrivals of the same soundwave will smear the transient part of any wave, so cutting down the amount of reflections will give you the purest, most intelligible sound.

First and foremost, empty venues have more acoustical reflections than a full one because most bodies absorb sound and stop reflections—hence the phrase: "It'll sound better when there are people in." This does depend on the space, of course.

Looking at the way the PA speakers have been placed in a room, you can see where the sound is going to be bouncing off walls and other untreated surfaces. Line arrays, for example, are sometimes hung quite close to walls; they have a fairly wide spread from the box. This means that a lot of the energy from the speakers is radiated toward walls, so a lot of energy is reflected off those walls. In addition, because of the nature of the hang, and because of the need to include the front row of the audience in your mix, you might find that a few of the boxes are being pointed straight toward the floor—and an empty floor reflects sound just as a wall does.

The problem you're going to encounter is that systems are set up in a way that work best for the person who set it up, such as the PA company, and not necessarily the people who are using it. Often, the PA company knows the best settings for the PA system, but not the venue—and without constant supervision and the ability to manipulate the speakers, you can end up with a pretty indistinct sound.

In situations such as these, how do you get a more distinctive sound? To begin, try turning off any speakers that are in the balconies or behind the mix position. (Make sure that you've listened to them and that they're reinforcing the sound from the main PA, and not adding unwanted frequencies to your mix in those places.) You can assume that the people in the room will soak up the ambient sound from these speakers before it has the chance to bounce back and affect what you're hearing. In addition, if there are places in the venue that aren't being used during the show, make sure that their speakers stay switched off.

A major consideration in every venue should be the sound that's being reflected back on stage. You should avoid reflected sound as much as possible, not only because those reflections will be picked up by the microphones, but also, and probably more importantly, so that the band isn't hearing beats and notes that are out of time. Many times the audience will soak up most of the reflected sound, but you will also come across low balconies that are fairly close to the stage, which can cause reflection (especially if the balcony barriers are made of glass). If you can, get some thick drapes to cover them up. If you're lucky, the solid barrier will be angled, and sound will be reflected away instead of into the audience or stage.

STAGE ACOUSTICS

As with everything audio, the most important element is the source sound. The better the quality of the signal, the better the product coming out the other end. Just as having a great preamp helps create a great sound, having a great source sound before the preamp helps even more. To ensure this, you need a good clean level on each mic—and the mics should have plenty of separation from each other, so no other sounds are heard through them. However, you are working on a stage and not in a studio—reflections on stage are common, and there is also a lot of noise on stage. To reduce this noise, you may need to EQ elements out of that sound. Stage acoustics become more and more important with the more mics you have on stage because the more mics you have, the more you're picking up ambient sounds in the background.

In addition to using different mic'ing techniques to get the separation you need, you can also help things along with some easy-to-use stage acoustic treatment. You can purchase all kinds of shapes and sizes of foam for this purpose. However, as a touring entity, it's easier to reduce the reflections on stage rather than in the room itself by using a simple bit of acoustic treatment that can be incorporated into the touring equipment or stage set. Remember, you don't want to eliminate the acoustics completely; rather, you merely want to reduce them or divert them.

The easiest thing to do is to get a backdrop made out of a nice thick, nonshiny material. The next easiest thing to do is to lay some flooring, such as a carpet on stage, which stops at least some reflections from bouncing off the floor. You can also use Marley flooring, which is a type of vinyl floor covering. It doesn't have the best acoustic properties, but because it's not a hard surface, it's better than nothing. (Marley flooring was originally laid down to help artists avoid slipping on stage, but it soon became apparent that it also helped with the sound.)

Another way to help cut stage reflections down is to use acoustic baffles. These are small partition walls that can be made up of a wooden frame that is stuffed full of absorbent materials (such as rock wool) and can then be covered up by a number of different coverings (plasterboard, cardboard, felt); this will absorb as many reflections as possible. These are a little more complicated because you need them to look like part of the set; structures that are about 7 feet tall and 3 feet wide are ideal. Place two stage left and right at the midpoint, turning them in slightly, and then another two on either side of where the drummer would normally sit, center back, also turned in slightly; this will cut down a lot of stage reflections. (For descriptions on how to make acoustic baffles and panels, the Internet is an excellent resource.) The biggest problem with baffles is that they take up a lot of stage space—but if you're having problems with lots of bleed on stage, they can be the way to go.

You might also want to think about covering the top of the stage in drapes, hanging directly over the band. This cuts out most of the reflections coming back from above the stage. In some places, especially old theaters, they used to drop bits of set in from the ceiling so that there was a lot of space above the stage; this can cause a few issues with ambient sound coming back down onto the stage. This also means that there is enough space up there to hang your drapes. It's not always practical, though, because not all venues have the same trusses or space from which to hang the drapes, but anything will help. You aren't going to run into problems all the time, but these are just some handy tips on trying to get that little bit extra out of your mix. You are basically making a studio on stage. Some bands won't go for this because it just wouldn't work for their setup, but if you are looking for a good-sounding show, then using these ideas could just be the way to go.

Now that we have stopped the stage from echoing sound all over the place, you must ensure that the microphones on the instruments are only picking up their own instruments. There will always be bleedthrough from one instrument into another instrument's mic, but there are a few things we can do acoustically that will help reduce this.

Stopping bleedthrough from different mics is all about thinking slightly outside the box; even simple ideas can be very effective. For example, the best way to stop sound moving from one place to another is to put a barrier in front of it. Think about this: A brass section can be fairly loud on stage, so its positioning can be crucial. When I toured with Amy Winehouse, the brass section was positioned just off center at the back of the band, angled in facing her. Though this might look good, the instruments were blowing straight into her ears and microphone, clouding her mic and what she was hearing through the monitors. To improve the definition, we ordered some custom-made Perspex screens that clipped onto the mic stand behind the mic and covered the bell of the horn. Having these screens helped clear up some of the ambient noise that was affecting the overall sound of her mic, and also cut down the level that was needed in the wedges because she wasn't getting blasted from the brass as much anymore.

The other major player in ambient noise is the drummer, who is normally positioned right behind the singer, and sometimes on a riser, making the cymbals right in line with the mic. The same method is useful here: Get a drum screen to block the noise from the kit (you'll find that the overall stage volume goes down as well). The only problem with putting up big screens is that they can really detract from the rest of the set, which doesn't always go over well with the management, record company, and/or artist. (They also require a lot of cleaning— there's nothing worse than looking at greasy fingerprints all over a screen.) Instead of getting full screens made, it is possible to get smaller screens made that just cover the cymbals at the front of the kit. At least this way you are still reducing the level of the high end that will be going down the vocal mic.

It's also important to reduce the amount of noise coming off stage. However, this can cause a problem for the artists on stage, so, as with anything live, it will be a compromise. One solution is to turn guitar amps around to face across the stage, or toward the back of the stage. If a guitar amp is on the floor with the guitarist standing right in front of it, angle the amp toward their head, and they won't need the volume as high.

The last thing you can do is improve the mic line, by keeping the speakers' sound out of the microphones. When setting up your stage, you should be aware in which direction the main PA speakers are pointing. Recall our discussion about the dispersion angles of speakers and trying to get the best coverage for the whole audience. Taking this into account, you may occasionally find that some speakers are facing across the front of the stage, so that the audience members in the front row can hear what's going on. If your main vocal mic is situated in this area, you might find it difficult to get the maximum out of the mic before it starts feeding back. I've always been a firm believer in keeping the mic as far back on stage as possible. (This can't always happen, as some singers like to be close to the audience—but do it if you can.) The only disadvantage to this positioning, as far as audio goes, is that usually you have the drums right behind the singer—so you might have a little more bleed from the drums. (However, if the drummer isn't too loud or is to one side, this is a good thing.) Another possible solution is to move the speakers farther from the mics and stage, and using highly directional mics. We'll look at the way mics work in Chapter 13, where you can see for yourself the different type of ways mics pick up sound.

So just to sum up everything here, the live stage acoustical environment can be a noisy place, and because of this, and as with everything live, it's a compromise. However, there are a lot of ways you can improve on-stage audio quality while still keeping the stage looking good. Always start off with the more subtle methods, but if you're still having problems, try more extreme measures. Remember that sometimes you need to dress them up a bit; after all, you aren't in a studio.

CONCAVE SHAPES

While we're on the subject of stage acoustics, let's talk a bit about concave shapes, or, more specifically, domed stages (see Figure 11.1). I've seen these more and more over recent years, and the acoustics of any form of domed stage are terrible. The concaved shape focuses the soundwaves on to a point—and, unfortunately, that point is usually where the singer is standing. Obviously, this is not good for keeping sound out of the microphone. This can be very hard to control, but there are a couple of things you can do. First, explain to your tour manager how it will ruin the whole show and that all the hard work the artist has done promoting and rehearsing for the show will be wasted unless the top of the stage isn't draped. If that doesn't work, try moving everything off center to avoid direct reflections into the mic.

Unfortunately, concave shapes are frequently used in amphitheaters (which are never perfect circles). Instead of the shape being overhead, it goes around the audience and does focus the sound for audience, but this was much more important when we didn't have amplified music. Nowadays, doing live rock shows from those venues causes all sorts of problems. The source point of sound is coming from the stage, and the venues themselves are built to amplify sound from that point. Normally in these environments, we hang a PA above where the acoustical optimized point should be; surely it would be better to work with the natural acoustics of the room and place the PA on stage pointing up and out into the room. I find that hanging the PA in the conventional way in these situations causes a lot of unpleasant summing in the back of the room.

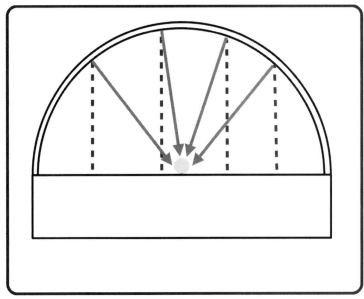

FIGURE 11.1
Due to the shape of a domed stage, the sound focuses back into the center of the stage, which can be exactly where the singer is standing.

The Royal Albert Hall in London is oval—except from the stage, where it is concave. The PA is hung in here, and the sound collects at the sides of the hall and is pushed toward the back, until both sides collect at the center back point, where it then sums together. This problem is a common occurrence in most venues shaped like this. (Also, in Albert Hall, the point at the back where all the sound sums together is where the mixing desk is—of course.) In this case, if we were to put the PA on the stage and point it up into the room, working with the acoustics rather than against them, we might get better results. However, we will have a problem with sight lines if we do that.

CONCLUSION

The mixing environment is an ever-changing world, and you will never come across a perfect environment when dealing with large-scale reproduction of music. A lot of factors affect the environment we work in, such as atmospheric changes like temperature, humidity, and wind direction. When doing a show outside, for example, you'll also have to deal with wind. In this case, using a PA system that has the power to push through mid- and top-range frequencies is important. Don't use line array systems. As much as they have the ability to not lose many decibels over a long distance, the actual power of the wavefront that is formed seems to be fairly weak, rendering the system pretty useless in anything over a slight breeze. Even if there is a lot of power in the system to get past the environmental factors, the level of the PA would be extremely loud. There are other environmental factors to consider in outdoor festivals as well, such as rain. At most European festivals it does have a habit of raining quite a lot. One minute the sun can be shining, and then the big black nasty rain clouds roll over the horizon and take a dump on your festival. At this point, the entire audience puts up their umbrellas or grabs their waterproof ponchos, which makes sound reflect off them, forcing the sound into the town just the other side of the hill and making the local residents start to complain. The sound has stopped being absorbed by soft people and is reflected off the shiny surfaces they have all bought with them. So what might have been a very slight rumble in the background has become a louder annoyance. It's not that there is any more volume; it's just that there are more surfaces for the sound to be reflected off.

Remember, your sound always has plenty of objects to bounce off of. Take a look at the venue and the surfaces that the venue has inside it. Be on the lookout for shiny metallic objects near your mix position, such as handrails or air conditioning ducts. I once had a polished brass plaque commemorating the opening of Folkstone's Leas Cliff Hall right behind my head. After getting some awful slapback from it, I eventually had to find some drapes to be thrown over it.

The PA is as important as the room, but the room doesn't change. You may therefore need to help it along with acoustic dampening, carefully placed Perspex screens, and graphic equalizers. A well-designed PA system should be able to fit into any room and should sound great in any room. If you're struggling with getting your sound right, this is an element that you need to resolve. Think carefully about how this can be done because you don't want to overcomplicate things for yourself.

CHAPTER 12
Tune Up

Following on from the wonderful world of acoustics, we land slap bang on the lap of tuning up your PA system. In this chapter we'll be looking at getting your PA ready for a soundcheck and the show.

PINKING

When you're all set up and ready to make noise, you first need to make sure that all of your speakers are working and that the right frequency responses are coming out of them. *Pink noise* is used for testing amplifiers and loudspeakers because it is a constant noise across the entire human hearing range. Pink noise power drops off 3 dB per octave, unlike white noise, which generates the same power at all frequencies. The pink noise is generated from the console and is then put through the PA to clearly hear what is coming out of where. You can mute each part of the PA system to hear the difference and then set your subs, mids, and highs to the desired level, also setting the level of any fills or delays you may have (although you might want to do this by ear later on). Pink noise is also a good way to find any faults within the system and also presents the best time to check that all your speakers are in polarity and in phase with each other. As the noise generated is constant and pumped through both the left and right sides of the PA, you will really be able to hear any phasing problems.

USING YOUR EARS

The most important thing to keep in mind here is to remember to use your ears! This may sound silly, but we are now in an age where computers can do everything for us, including how to set up a PA. Just because they *can* do it doesn't mean you should *let* them do it—as it doesn't always work out. I once turned up to a gig in Mexico City to find a flown PA system hung very wide on either side of the stage and pointing to the very top back corners of the room. (And when I say "room," I mean a 4,000-capacity concrete indoor amphitheater.) The FOH console was located halfway up the slope on one of the main walkways. As I was standing there looking at the stage and then the PA in relation to the stage, I noticed that it didn't even point at me—and I was about 70 feet from the stage. After playing some music and walking around, I was truly shocked: The entire front section of the audience couldn't hear direct audio from the main PA (except for infills, which would cover the first two rows). When I asked why the PA was pointing to the top back corners of the room, the answer made me cringe: "That's what the computer said to do."

You can certainly use the software provided with the PA to get the right speaker angles, but if you hang a PA and it doesn't sound right, it *isn't* right—no matter what the computer says. This is an important point that can truly be applied to everything: If it doesn't sound right, it isn't right.

USING EQS
What Is EQ?

Equalization (EQ) is the manipulation of the frequency response of a sound, which changes its perceived tonal balance. It is a fantastic tool for adding depth and clarity to your mix, and it is not merely just a tool for removing feedback.

When you're talking about manipulation of tone of audio with rooms, PA systems, guitars, drums, and basically anything else you listen to, you're talking about EQ But when you're asked what kind of EQ you've put on the guitar, the response might be "none." In this context, "EQ" refers to anything you do on the console to make the guitar sound like it does.

Room EQ is a correction to the natural frequency response of the room, which you can change by changing the way the speaker sounds to compensate for the sound of the room. Specifically, the *room EQ* would be the curve (frequency response) you've set on your graphic equalizer (GEQ), also known as a system EQ (because you are EQing the system to the room).

How EQ Works

When the first equalizers (EQs) were made, they were designed to pass a signal through capacitors and inductors. Those components acted as passive filters to change the frequency response of the incoming signal. Depending on the type and size of the components, you were able to select different frequencies to adjust their relative levels. From here, all different types of EQ were developed.

Digital EQs work on the same principle, but instead of passing current through capacitors and inductors, they use digital signal processing (DSP).

When you start to look at nonfixed EQs (parametric EQs), you can see how the relationship between the EQ and the phase is changed. With wider bandwidths, you are affecting more frequencies, so more of the frequency range is affected. With narrower bandwidths, fewer frequencies are affected, but you get a greater phase shift. Harmonics and subharmonics are affected by their relationship with the phase-shifted frequency, so by using EQ in the midrange, you could also be affecting the sound of the lows and the highs.

With the development of newer digital technologies, we are beginning to see things like linear and nonlinear phase response, minimum phase, complementary phase shift, and constant Q. All these EQs use some kind of phase adjustment to work, then try and readjust the phase shift to be as close to flat as possible (or at least minimal).

Phase response is extremely important within what you're hearing, to both audible and inaudible frequencies. Research in the 1970s uncovered extremely convincing evidence that having a frequency response up to 100 kHz is important, but microphones and speakers only go up to about 20 kHz. We can't actually hear frequencies that high—it's the phase response in the audio band that matters. Near the frequency where the device you're using starts to roll off, the phase shift increases. Some might say that this is why hearing a live acoustic concert without any amplification provokes more emotion and has some sort of deeper meaning than listening to the same concert in a recorded format and playing it back in your living room. And as a little side note, the sound of digital audio is related to the phase response of the filters in the A/D, D/A converters.

The very nature of EQ is changing the shape of the phase response of the equalizer, and the narrower the band, the more phase shift there is. This can cause a problem, though—because, as you know, when you have two signals that are 180 degrees out of phase and mixed together at equal levels, you won't hear them at all. The importance of the relationship between EQ and phase doesn't mean that you shouldn't use any EQ; it just means that you should use as little as possible, and where and how you equalize is important, like choosing good-sounding microphones instead of using desk EQ to get the sound you want.

Room Response

Before we look at the actual process of EQing, we should take another quick look at what the room gives back to us in the way of room acoustics affecting the sound from the PA speakers. As we said in the last chapter (11) about acoustics, when you put a sound into a room it talks back to you as a reflected sound. It's this frequency response caused by room relections that we want to look at. Because EQ and phase are intertwined, a room will not only have a frequency response, but it'll have a phase response as well. Along with many other factors, phase is also responsible for some of the room's EQ; how much depends on the room (angles of speaker boxes, room resonance, standing waves, and so on). Phase becomes another interesting factor (think about all the tiny delayed responses bouncing off the walls, all of the same waveform) when starting to EQ your room. So there are two arguments here: Using a graphic EQ can change the phase response dramatically, so you should use as little EQ as possible. But if the room's phase response isn't flat, by using a graphic EQ you could realign the phase. The point is to use your ears and walk the room thoroughly, listening to the way the entire frequency range sits in the room. You will never get a perfect sound everywhere, but you can try to get a great average sound everywhere. The most important factor is making sure all the individual elements of the PA sound good at the start rather than compensating for a poorer sound quality using the graphic EQ.

To measure a room's frequency response, you use a *real-time analyzer* (RTA), such as EAWs Smaart software. It gives you a visual indication of the speaker-and-room's frequency response and phase response. The problem here, though, is that rooms have different responses depending on where you're standing. Looking at an RTA measurement at one location can really distort your view of the room as a whole. But the best types of analyzer of all are those big floppy things stuck to the side of your head. You know it will sound different in different parts of the room, so you can compensate for that by walking the room and understanding how it sounds at the place where the majority of people are standing compared to your mix position. The RTA only shows what a microphone is picking up in one location.

The other thing you need to think about when using an RTA is that you don't want to have a flat frequency response. We don't hear with a flat response; flat is boring. We like excitement! A measurement microphone does not pick up sound as our ears do, so a flat measured response always sounds too bright (strong in the highs). The measured frequency response should roll off gradually above about 1 kHz to sound subjectively "flat."

In addition, trying to EQ for a flat frequency response probably destroys the phase response of the speakers/room. But in my experience, it's better to have a flat phase response than a flat frequency response. Having a flat frequency response doesn't mean you have a flat phase response, but having a flatter phase response will probably lead to you having a better show.

Processes

Let's look at the processes engineers use to EQ a system. Some people prefer to *voice out* the PA, which means that the engineer EQs a system simply by listening to sounds they make into a familiar microphone, such as various and random clicks, clucks, and the words "one" and "two." (Using the words "one" and "two" actually has a purpose; other than helping a novice who is speaking into a microphone look like a pro, the word one contains lower and muddier tones in your voice and the "t" sound in two contains highs.) The idea behind tuning the PA system to a microphone is that you know your voice and how the microphone should sound on a well-tuned system. However, due to the proximity of the microphone, your voice contains low mids that might be emphasized. Low mids give tone to your bass guitar, drive to your guitar, and power to your vocal; if you remove them, you're going to have a hard time getting any depth to your mix. You might get a great vocal sound, but your voice isn't able to re-create the high and low frequencies you want and need throughout the whole frequency range in your mix. You may start by taking these frequencies out, but in the end you'll be adding them back in. Again, this is an important lesson: You must know when to EQ and when not to EQ. The other problem in using your own voice to EQ a system is that it's your own voice, and you aren't the one who's going to be on stage singing. Also, there is nothing more annoying for the rest of the crew than listening to you shout through an SM58 for 20 minutes or so.

Another EQing process is called *ringing out*; the idea here is to turn the microphone up until it starts feeding back, then find that frequency on the graphic EQ, and pull it out until the frequency stops feeding back, then repeat. This procedure is something that a lot of people use on monitor systems to get as much volume out of the microphone as possible. The rule of thumb in this process is that, if you're pulling out anything more than six frequencies, you're doing something wrong. That might be the case, but, frankly, if you're using this process, you're doing something wrong in the first place. It may seem to work okay on monitors, but you could still be clouding the monitor mix with unwanted low mid that, if just taken out, would clear everything up and you wouldn't have to have the monitors so loud. And at FOH, you should *never* use this process. We want our mix to sound great, not just loud. Loud doesn't get you anywhere. This procedure, whether it is being used on monitors or the front of house, is extremely annoying for anyone else trying to work in the room and is considered by most industry professionals to be extremely amateurish. Instead, you should understand what the frequencies do in terms of how boosting and cutting each frequency on the graphic affects the overall sound, and you should use them to your advantage by understanding that using a combination of cuts or boosts will help your sound in the end. A microphone should have pretty good gain before feedback, so feedback shouldn't be much of an issue. If it is, you need to be looking at how the speakers are placed in the room or on the stage, and whether the microphone is close enough to its sound source and is not defective. Do not sacrifice the sound of the system just for a little more volume.

I prefer to get a piece of music and play this through the PA. This way you'll have the frequency range and dynamic range that you are looking for through the show, and you'll have the ability to walk the room a lot easier than trailing a mic cable after yourself. The music recording must be something you know well, and also something that is well produced. Some engineers use one particular track, whereas others have a number of tracks from which they choose. It's really important that you know these tracks inside out, including the subtle parts in the background, so that when they are played you can hear them. And if they aren't there, you know you need to emphasize them with EQ. Use a piece of music that has a bit of punch, some low frequencies, a good vocal, some highs, and—most of all—space. A lot of CDs these days have been re-mastered, with a lot of the levels turned up. You're losing a lot of dynamics in these CDs, so try to avoid the latest releases, anything that is re-mastered, or a greatest hits CD. When using a CD, make sure you don't have any EQ on the channel and that you don't have any filters, inserts, or routing to anywhere other than your main mix send. Make sure you have a good gain trim setting and then push the fader up.

Never, ever, under any circumstance, use an iPod or any other form of personal digital music player, including a CD that has MP3s burned onto it. Get yourself the original CD and use that. The DAC (Digital to Analogue Converter) on CD players is far superior to what you find on your personal digital music players, even if the sample rates and bit depth of your music are at their highest. It's the outputs that count, and you'll hear a massive difference. (Apparently you aren't supposed to be able to hear data compression with an MP3 file encoded with a very high bps rate, but this is just not true.) Of course, using recorded music won't give you the exact same response as the band about to play through the PA, but it will give you a decent indication of how the room is responding to the frequencies you're putting into it.

You might find that some engineers have one track, their signature track. Personally, I use a combination of different things from Rammstein's "Reise Reise," Chaka Khan's "Ain't Nobody," Venus Hum's "Heaven" to Genesis' "No Son of Mine"—it really just depends on how I'm feeling. Music is all about a feeling; you need to have that feeling to translate it and the emotions to understand it. It's a base level thing.

With all that in mind, before you jump on the graphic and start pulling the graphic EQ's faders up and down, you need to think about what you are hearing in the room when the reference CD plays.

Graphic Equalizers (GEQs)

When attempting to calm the frequency response in any room, the first thing some engineers often tend to do is carve huge chunks out the frequency response with the graphic. However, this isn't really the best way to go. You should think of graphics as creating an audioscape for your mix to sit on; they bring more of an artist's perspective on what the engineer thinks the room should sound like. Graphics themselves are actually pretty useless at sorting out anomalies in a room because they can create anomalies of their own. They should therefore be used to make the room sound how you like it, and let the more advanced system parametric graphic take care of the room anomalies. One reason for this is the way the filters work. The filters within a graphic are notch or peak filters that cover a wider band than their center frequency; they interact with the other frequencies around them, as well as the frequency selected. When you want to pull out two frequencies next to each other, it tends to cause a kind of ripple effect in the frequency response, where the filters don't meet at the same gain boost or cut, so the frequency where the two filters meet has more or less gain than the center frequency. In addition, not all the boosts and their equal cut hold the same bandwidth. You might find that when you boost a frequency, the bandwidth is wider than when you cut that same frequency.

The other reason GEQs are pretty useless is that they are split up into 31 1/3 octave fixed bands (you aren't able to change the frequency), and the only control you have over a graphic is how much cut or boost is applied to that frequency—which makes it pretty difficult to "tune" the room on a graphic. Parametric EQs,

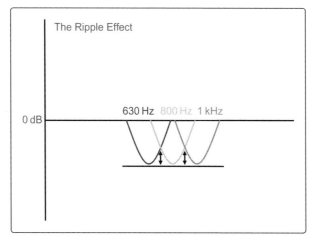

FIGURE 12.1
The Ripple Effect—This is caused when two frequencies next to each other are pulled out on the GEQ.

on the other hand, have control over all three parameters, so you can easily use one of these to get in between the frequencies and pop out problem areas.

With the advance of newer digitally controlled GEQs such as the Lake Contour, you can now fine-tune your system and the room. This kind of GEQ is more like a parametric EQ, but in the graphic type format. The graphic EQ gets its name from the fact that it is graphical, naturally, so we can call our parametric visual EQ a *paragraph*. This type of EQ is far better for tuning the anomalies within a room than the conventional GEQ.

Here's the way to think about it: 31 band GEQs are better for creative purposes. They can't be used as an actual tool for system control because of the types of filter they have on each control. Wider bandwidth controls are generally used more for artistic license; they have smoother responses and interact more with the frequencies around them. ParaGraphs should be seen as a tool for adapting the system to the type of room you're in, and for cutting or adding just those frequencies that need to be adapted. Whichever way you look at it, you're going to be changing some kind of phase to remove some room issues.

One last point before we move on: With any digital console, always try and get a *grab graphic*, which is a physical GEQ that is placed in a rack beside you (not just the onboard graphic of the console), is normally set to a flat response, and is used on the left and right speaker stacks of the system. This means that if you have any problem frequencies, you can easily get to them rather than having to flick into a mode where you either lose all your faders (because the digital console puts the GEQ control over your channels) or you end up having to look at the screen to find out to find out where the cursor is (so you can move it across to the problem area). Yes, this does defeat the point of having a digital graphic inside the console, but, frankly, the show only happens once, and if you're too busy trying to find 2.5 kHz on a screen, there's something seriously wrong.

THE ART OF SYSTEM EQUALIZATION

In this section, we'll look at the actual process of equalization, and we make the assumption that you have a system GEQ and a rack-mountable GEQ. On small PA systems you will probably only have GEQ, but you will find that bigger systems have a GEQ at FOH that you can adjust, and also a system GEQ incorporated into the crossovers. These types of GEQs are only accessed by either a computer or a button pressing combination on the front of the crossover. You might find some kind of computer screen that is connected to the crossover, so the system tech or in-house engineer can set the system up the way they want it, and then all the other engineers who come in and out of the venue use the graphic at FOH. A lot of the time you'll find that the system EQ curve is locked, whereas other times the in-house engineers are happy to turn it off. Remember that, when you're in this situation, you're in danger of double EQing.

Graphic EQ is used to tune the frequency response of the speaker to the room so that it sounds accurate or hi-fi for all sound sources. Desk EQ is used as a creative production tool to alter the tonal balance of each individual instrument and vocals.

When EQing, look at where you're positioned and how that will affect the sound, as well as where the audience will be. The biggest thing to remember at this point is that you are just looking to remove any anomalie, that is, any frequency or frequencies that stand out more than they should. Remember, though, that when EQing the loudspeakers, you're creating phase and frequency shifts over the entire mix. Applying a 400-Hz cut over the whole system means that all your instruments will be cut at 400 Hz. If the tone from your bass guitar comes from there, or that is where one of your toms is tuned to, you'll have to boost those frequencies to bring that out again. Changing the phase over the whole mix could be a potentially bad thing because your audio ends up sounding like it's "tiny" rather than being wide and open. Individually, though, EQ changes on the channels can and will enhance the mix. The following question arises: Is there too much frequency emphasis in the room or in the mix? You need to really listen to what you're doing. Trial and error helps, but experience is the best teacher. Once you understand the listening part, everything else is fairly natural.

As you're listening to your music through the PA, before you touch any form of EQ, you need to get out from behind the mixing console and go for a walk. There might be frequencies that are more prominent in other parts of the venue than others, and you also need to listen to how the rest of the venue sounds in relation to where you are. Keep in mind the differences between where you are and the rest of the venue. If you stand in the middle of the room and it sounds great, then leave it; most of the audience will be standing there. You're listening for frequencies that jump out at you; these will present themselves as kind of loud notes being played. You're also listening for frequencies that you *should* hear; this is why you need a track that you know well. What you're listening for translates to all EQ, and it too comes down to personal taste. Once you've decided which frequencies are sticking out, head over to your GEQ and remove them. One thing to keep in mind while going through this process is that not all PAs can reproduce everything equally well. If you are listening for some subtle percussion in the music you are playing and it seems to be stuck very far in the back ground, you might try and EQ so that it stands out a bit more. You are in danger here of adding too much EQ that could swamp the rest of the mix. Listen to how those frequencies sit in the room, and your mix, but make a mental note of what you have done and review it later.

Don't try and do much EQing on the main system EQ. The room will respond differently when there are people there, so understand what you're pulling out. If you end up pulling out a lot of highs, be aware that you might have to push them back in later on. It's far easier to do this on a graphic than it is on the EQ hidden away within the crossover, using a computer screen and magic pen that moves things around.

As we've said all along: listen. Just because you have a graphic right in front of you doesn't mean you have to use it.

As a touring engineer, I've never been afraid of getting my hands dirty with a graphic. Once, at the House of Blues in Houston, I raised the faders up on the CD I was using to EQ the system, and I heard what can only be described as a bloody awful sound coming out of the speakers. I turned the CD down, checked that the EQ wasn't in on the channel, no high pass (HP) or low pass (LP) filters were in, checked the gain, and even checked the CD through the headphones to make sure the sound was good on my end. When I couldn't figure out what was wrong, I asked the in-house engineer, and I was informed that the system always sounded like this. To try and fix it, I played around with the graphic. When I got something that was at least remotely close to what I wanted ("remotely" being the key word), the screen looked like I'd been playing a game of Battleship. I asked if this was a typical EQ curve for engineers to put on the system, and his response was: "I've seen some pretty extreme EQs in here." The system was installed by a PA company, and then the in-house guys were locked out of any proper control over the system. The system probably would have been set up with the standard factory settings, and placed in the room when no one was in there. It would have then had a general system EQ placed over the whole thing, the crossover points set, and would have been locked and left for the in-house techs to get along with. This meant that they couldn't improve on the sound as more and more shows were coming through—which means if that is an engineer's first experience of an Adamson PA, he'll never want to use it again.

The point is that sometimes systems just sound bad, so don't be afraid to jump in and push around a few GEQ faders to get the sound you want. Because of the way the graphics are split up, you only have 1/3 octave steps (1/3 octave bandwidth for each fader), so don't be surprised if you end up cutting one frequency and boosting the next to try and find an in-between. It's not the most efficient way of doing things, but sometimes you can get the sound you need, and as with anything in live audio, there is always a compromise.

Creating an Audioscape

The canvas in which you're going to place your audioscape starts here, with EQ; you're building a background for all your other EQs to sit on. Remember, the EQ you have on the channels works in conjunction with the EQ over the whole system. This is where you can make an instrument stand out or, alternatively, fade into the background. For example, you may drop out some low mid and then find that when you work through your channel EQs, you have to put low mid back into the drums, bass guitar, guitar, and so on, all because it makes your vocal less muddy. In this case, the better thing would be to just remove it from the vocal channel, rather than the whole system, keeping the phase alignment of the PA a little more intact. When you use the channel EQ and compression on a console, you work in a way that helps you control your mix.

If only the vocal sounds boomy, don't use the house graphic EQ to improve the vocal tone. Use the desk's channel EQ on the vocal only.

One of the first EQ rules I was taught was to not add gain to the EQ. So, these days, that's exactly what I do. The reason for this is that it might add distortion if the signal is already near clipping, which it should not be doing. However, distortion isn't always a bad thing—too *much* distortion is a bad thing. Adding a touch of it to an instrument can implement drive and a little rawness (though this is more of a channel issue than a system EQ issue).

You have a lot of power at your fingertips. You can make a sad song sound powerful and full of emotion, or weak and dull. The trick is to work with your GEQ and your console EQ together. Be aware of when to EQ in the house and when to EQ on the console. I get into that a little later, but it's really simple if you just follow two simple rules: listen and think. Listen to where the frequencies are coming from; if you can hear the 250-Hz woolliness in the guitar, and not in the bass, then chances are it's just in the guitar. Think about where you're going to be pulling those frequencies out. If by removing 250 Hz you lose depth in your mix, then you need to think about where else you can de-mud the audio.

Fills and Delays

Before applying any EQ, time alignment should have been done for all your fills and delays because unaligned speakers can affect the frequency response. Don't forget to check how the fills and delays sound, but keep them off while listening to the main system. This way they won't be clouding what you're hearing. Once the main PA has been EQed to your liking, you can turn your fills and delays on.

Depending on the size of the stage, keep as much out of the infills as possible. Put mainly vocals through these, as you will probably get a lot of the audio from instruments back from the stage that will cover the front few rows of the audience. As far as the infills are concerned, they'll probably just need a little bit of vocal boost, so you can EQ the infills for what's best suited to the vocal rather than what's best for the whole mix. This is just to reinforce what is coming out of the main PA, so there is no need to have it extremely loud. In fact, you shouldn't really be able to hear where the sound is coming from. You might find that you want to add a few of the main melody parts to the infills as well.

Your outfills and delays are receiving the same mix as the main PA loudspeakers. The trick here is, once again, to simply reinforce what's coming out of the main PA system. If you just need a few more highs at the back, then just put the high in the delays. You don't want to cloud the delays too much. The job of the main PA loudspeakers is to give you your sound pressure level; the job of the delays is simply to reinforce the parts that can't be heard.

Because your outfills tend to cover parts of the audience that have next to no direct signal from the PA, you're going to have to put the entire mix through them. Try to EQ these for the whole mix. Chances are you won't need any sub-frequencies in here because you'll be getting enough sub from the main PA. The level will really just depend on the size of the speakers you have. If they're just covering a small amount of the audience, then turning them up too loud is going to give you problems at the point where the main speakers take over from the outfills.

The best way to have any of this set up is so you can't hear any individual speakers take over from one another. They should all be working together, and all should sound like they're coming from the same place. As long as you have the correct delay settings for all your speakers, and the EQ is pretty accurate, this should be a fairly easy task to accomplish. Adding the right amount of delay to the outfills channels makes the listener localize the sound on stage rather than at the nearest outfill speaker.

You will probably not be able to hear any of the fills or delays from your mix position, so it's very important to make sure they sound good at the start. If possible, always have an EQ on these sends. The other thing to bear in mind is that, if you're going into a venue with its own PA, these sends and EQ might already be set up. Before changing anything, have a walk around and listen to them. You probably won't need to change anything, if not that much at all.

It is always a good idea to turn all your fills and delays off for most of the soundcheck; this ensures that you're getting sound directly from the main PA rather than ambient sound coming from other sources around the venue.

Feedback

One of the questions that is always asked is how to eliminate feedback. Obviously, eliminating feedback is extremely important, but not always easy. Many factors are involved in causing the problem, so there are many solutions for getting rid of it. The first thing most people do is to jump on the graphic and start hacking frequencies out. This gets rid of the frequency that is feeding back, but it could also destroy vital clarity.

You may see a device called a *feedback exterminator* or *suppressor*. Do not ever use it. It might get rid of the feedback, but it also might also get rid of the vocal.

This device is a tool for small pub bands that set up their own PA systems, not for professional engineers. If you ever see one in a system, you should seriously question the person who installed it. Feedback exterminators detect the frequency that is feeding back and reduce that frequency in level until it stops feeding back. I've never come across one that works well; as soon as a frequency starts to jump out, it pulls it out, even though it might not be feedback.

In the next chapter we'll look at different polar patterns of microphones that will help to increase the *gain before feedback* level. But some other things to think about if you are having problems with feedback are moving wedges or FOH speakers if they are interfering with your mics too much, or placing your mic closers to the sound source.

Not only is feedback very annoying, it is also bloody dangerous. The first thing you should be aware of is *what* has the potential to feed back. There is one really good tool for working out what these potentials are: your eyes. Look at your mic positions to see which mics are closest to the wedges, then think about how much input gain these mics usually need. The other thing to look for is who or what is placed in front of the mic. This is related to changing the polar response of the mic, which we'll go into in more detail in the next chapter. But for now just understand that when you place an object in front of a mic, that object reflects sound into the front of the mic. If someone were to talk into a microphone, instead of just picking up the sound in front of the mic, it's now picking up sound from around the sides and rear, and this might be where your wedges are. Also, chances are you'll either be touching, or have just touched, the channel that starts feeding back. If something starts feeding back, you need to be quick; mute the channel that's going off, pull the fader down, and start to push it back up until it's more controllable. Then find the frequency either on the channel EQ or your GEQ and pull it out (just remember what we have just talked about, and think about *where* you need to EQ). Really, the best way to get rid of feedback is to use your ears and listen to what's going on. Professional equipment is used all the time without any problems at all, so if you're finding yourself struggling to contain feedback, then maybe there is something wrong with a piece of equipment or how it has been set up.

CHAPTER 13
Stage Setup

At this point, it's time to join the crew on stage. The stage set is a big deal because it's what you're presenting to the audience. At this point, it will be a hive of action: The monitor engineer will be running out cables to his monitors, and the in-house engineers will be looking around for instructions, as long as they haven't all gone to the crew room for tea and biscuits. The backline techs will be busy building all the backline and other paraphernalia that has to be set up. You'll probably be shouting at the lighting techs because they've decided to put MAC500s on flightcases that are sitting right on top of all your mic cables, and the production manager will be nowhere to be seen. Somehow, though, the whole team will come together and make the stage look wonderful.

The stage plan and channel list you've painstakingly put together will come in very handy now; they help your local sound techs and crew place all the articles you've requested on stage and in the right place. All your risers will be in position, your satellite mic boxes will be run out, and hopefully you'll have microphones on stands next to the instruments you want them on, ready for you to place them. (Of course, this isn't always the case.) Try to keep everything as neat and tidy as you can; that way, everything will look professional, you'll know where everything is going, and, most importantly, when the lights go out for showtime, no one is going to be tripping over anything.

From your point of view as an audio engineer, the main element of the stage setup is your microphones. There are many, many different kinds of microphones, and they work according to three different transducer principles—but all of them revolve around the principle of sound vibrations being picked up and transformed into electrical signals to be sent on their merry way down the mic line to be mixed for your listening pleasure.

You have a world of choice when it comes to microphones; there are many different kinds from many different manufacturers, each with its own unique properties. Once you've experimented with a few different mics and a few different positions, you'll soon learn which ones work best for you.

A lot of the time, you'll use specific mics for specific purposes: vocal mics, instrument mics, and the like. However, it's always good to try out a few different combinations. For example, the Shure SM58 is an industry-standard vocal mic, but you can also use it on a guitar, bass, snare, or kick drum. Similarly, the SM57 works well for pretty much any stage application. These mics are great because they work on anything and make the instruments sound decent. They aren't fantastic, but you can still get a great mix out of just using 57s and 58s. Another one of my favorite mics is the Sennheiser 509 or 409; technically, it isn't "supposed" to be used on a vocal, but I like the effect it has. You can't really use them at a live show, though, because they end up covering the vocalist's face. They look a little like old-school mains-powered shavers, but the vocal sound is really cool.

I'm also a big fan of the Audio Technica AT4050 (one of the best mics on the planet, as far as I'm concerned). They sound amazing on guitars (and they do well on overheads as well, but they can give you so much more than that). The one major problem with these kinds of mic is that they sound *too* good: If the most expensive mic in the mic box is on your guitar, that guitar is going to stick out. The signal might be too crisp and clean, so when you put it next to your SM58 on your main vocal (with a price difference of nearly $500), it just won't sound right. You need to select the right mic for the right job. If you have a budget for good quality mics, then great; if not, choose wisely.

If you don't have the time to experiment, always go with what you know. If someone suggests something, use his or her advice, but also don't be afraid to switch back to what you know. Mics are a very unique thing, and what works for one person might not work for everyone else.

DYNAMIC AND CONDENSER MICROPHONES

In the live industry, it's really only practical to use two types of mics: dynamic and condenser. (Other types of mics, such as ribbon mics, are simply impractical; they're far too delicate to take on tour, and they also are usually on the expensive side.) Still, the choice of mics within the dynamic and condenser ranges is huge, so don't think you won't be able to find anything good.

The most common mics you'll find in any live setup are moving coil microphones, which are more commonly known as *dynamic mics*. These tend to be rugged, reliable, and generally good, hardworking microphones. Their price range is quite large, from only a few tens to a few hundred. You'll find that the sonic character of these mics tends to be a bit more colored than condenser mics, and because of this you're sure to find a mic for all occasions. If you're only presented with moving coil microphones, you can always make your band sound good, although cymbals and acoustic instruments really benefit from the better transient response and high-end clarity of condenser mics.

Condenser mics tend to sound very clean and have an extended high-frequency response. They're used on things like hi hats and as overheads, but their use goes far beyond just that.

Below we'll have a look at the difference between these two types of microphone in more detail.

Dynamic Microphones

Dynamic mics work exactly the same way as speakers, but in reverse: The diaphragm of the mic picks up soundwaves, and an attached voice coil suspended in a magnetic field converts the diaphragm vibrations to electrical signals, which are then sent to your mixing console for mixing. Within the mic, the diaphragm, which is a thin membrane, is glued to a copper-wire coil, just like a speaker. As soundwaves hit the diaphragm, the diaphragm moves, causing the copper coil to move as well. Very near the coil are the two poles of a magnet, and the moving of the coil through the magnetic field of the magnet creates the electrical signal. This is why they are also called *moving coil* microphones. Just like speakers, microphones have a transient response; dynamic mics tend to have a slower transient response than condenser mics because the diaphragm and voice coil in a dynamic mic have more mass to move than the light, thin diaphragm in a condenser microphone.

Most dynamic mics can be turned up louder than condenser mics before feedback occurs.

Condenser Microphones

Condenser mics need a power source to power an active circuit inside the microphone. A thin, metal-coated plastic diaphragm is mounted very close to a metallic disk called a backplate. The diaphragm and backplate are charged by the power source to form two plates of a capacitor. Or the backplate is coated with a permanently charged electrical material. When soundwaves vibrate the diaphragm, that varies the capacitance and generates an electrical signal (varying voltage).

Because of their design, condenser mics give far more accurate reproduction of the incoming waveform with less color. This is why a lot of producers use them as vocal mics in the studio. Because they tend to have a wider pickup pattern than most dynamic mics, you can have trouble using them as vocal mics in a live environment; you might just not be able to get enough level out of them before they start feeding back. In addition, you can pick up a lot of background noise (leakage).

Some condenser or dynamic models tend to sound a little harsh, depending on how the PA is EQed and on the voice of the vocalist. This is a good lesson: You must match up mics with voices rather than just always relying on good quality mics.

When I first started working with La Roux, Elly had a KMS 104 by Neumann. This is a very good quality mic and sounds great, but Elly's vocal range is really strong around 1 k, and this mic just brings it out more, leaving the rest of the frequencies in her voice to suffer a little bit. The previous engineer had insisted that this was a brilliant mic, and he wasn't wrong—but when I persuaded them to change to a Sennheiser E945, the difference was amazing. The body came back into her voice, she sounded fuller, and she could hear the massive difference between the two mics in her IEMs.

FIGURE 13.1
La Roux's Elly Jackson with her radio mic the Sennheiser 500-945. This is the radio mic version on the hardwired E945 mic.

PHANTOM POWER

To power condenser microphones, you need a convenient power source. These come in a couple of different forms: Some mics take batteries, while others are powered by a small box that plugs into the mic—but by far the easiest and most convenient way to power a mic is through the mic cable itself. After all, it's already running to the mic—so why not send power down it? +48 volts is applied equally to pins 2 and 3 on the XLR; that voltage is relative to pin 1, which is your ground. (You might think that sending this kind of voltage down a mic cable would mess with the audio signal, but it doesn't.) These days it's very rare that you come across condenser mics that need their own dedicated power supply; a circuit within your mixer will usually provide the power at each mic's XLR connector. (However, if you're plugging a mic into an audio interface to use with a computer, you might need to use a separate supply; not all interfaces have phantom power capability.)

Phantom power is a DC voltage, and microphones draw only a few milliamps of current from the phantom supply. Different mics draw slightly different currents. If the phantom power supply doesn't reach the intended operating current for that mic, you will still get signal—but probably with reduced signal level or some distortion. You'll hear this a lot with some active DI boxes (which we discuss in more detail later in this chapter).

Handling Noise

Occasionally, you'll start to mix and hear sounds that you haven't noticed before—you can hear finger tapping and big clunks as the mic is put in the clip on the mic stand. This is called *handling noise*. With a decent microphone, there shouldn't be much of this (although you can never completely eradicate it). When it becomes noticeable, the mic's internal shock mount (if any) has broken. If this happens, you can send the mic back to the manufacturer for a quick repair.

SONIC CHARACTER OF A MICROPHONE

As we have just looked at how dynamic and condenser mics work, you should understand that because of the way they work they give us different *sonic characters* (which is the perception of the mic's *frequency response*). Many factors affect the sonic character of a microphone, such as whether it is a dynamic or condenser mic, and the materials that make up the diaphragms that pick up the sound waves. Each microphone is different, and even microphones of the same make and model can have different sonic characters (usually due to being used more or damaged in some way). Eventually, due to age, they can start to change their sound, as bits of dust and dirt get into the internal workings. You may find that you find a mic that sounds really good on something; it just adds a little extra niceness to what you are mic'ing up, but when you try another mic of the same make and model it sounds different.

There are a few other things that make up the different sonic character of each mic as well as just normal deterioration. Next we'll have a look at some additional factors that affect individual microphones' sonic characters.

Proximity Effect

When you move closer to a directional mic, the bass response increases more and more. This is what we call the *proximity effect*. Mics that have an omnipolar pattern (see later in this chapter about polar patterns) do not have a proximity effect.

Critical Distance

Typically, with close micing you won't really come across the *critical distance* too often. This is the distance from the microphone where the source sound and the reflected sound (from the same source) are at the same level. If the microphone is at a critical distance, the ambient noise and the direct signal will be at the same level. If the mic is then moved closer to or further away from the sound source, this will tip the balance to how much source sound or ambient sound there is.

POLAR RESPONSE

Whenever you pick up a mic, you must know which way to point it, and you need to understand what the mic could pick up from the surrounding environment.

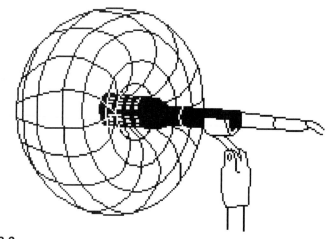

FIGURE 13.2
A 3-D image of how the cardioid polar patterns work.

Each mic has its own *polar pattern, pickup,* or *polar response,* which is the mic's sensitivity vs. the angle of the incoming sound.

Most microphones used by live sound engineers have a *cardioid* polar response. Cardioid patterns are best known by their heart-shaped pattern (Figure 13.3), and the name comes from the Greek word *Kardia,* meaning heart. This pattern represents the area from which the microphone picks up sound in various degrees. In the case of the cardioid pattern (Figures 13.2 and 13.3), you can see that most of the sound is picked up from the front and a little less toward the sides (−6 dB at the sides). There are a few different types of unidirectional (cardioid) patterns (Figure 13.3) that we use on a regular basis: cardioid, supercardioid, and hypercardiod. (There are also ultra and sub cardioids patterns, but these aren't used very commonly.) Another type of polar response is the bidirectional, or otherwise known as figure 8. Microphones with this type of response pick up sounds from both sides of the mic. There are also omnidirectional polar pattern mics, which pick up sound from all around.

When close micing (see mic placement in this chapter below) instruments on stage, it's a good idea to stick to using cardioid-type patterns because there is less of a chance of picking up ambient noise and spill from other sources that will cloud what you're doing. Polar patterns that pick up the room are great in studios, but when doing a live show, everyone is already in the room—so that can be a little bit pointless.

Some mics have different polar patterns that you can select, which gives you a lot of options and can enable you to use good quality studio mics in a live environment. The major advantage of having these types of mics in your show is that they are precision mics and can really enhance the sound of what you are mic'ing up. But don't forget about choosing your mics wisely, for using a

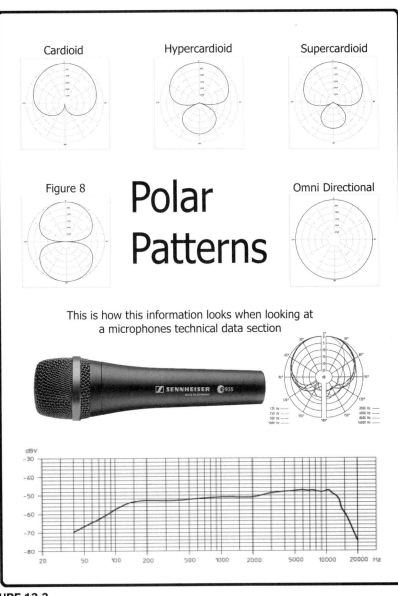

FIGURE 13.3
Cardioid, hypercardioid, supercardioid, figure 8, and omni directional polar patterns.

mic that is far better than the other ones in your mic box won't help your mix sit together very well.

Changing the Polar Response

When you introduce anything in front of a microphone, you are changing its polar response. Regardless of the type of object—a speaker, drum, glockenspiel, someone singing, and so on—this has basically the same effect of changing the polar response, but to varying degrees (depending on how far away the mic is from the source).

From time to time, you'll see engineers cup their hands over the microphone. This causes feedback, so don't do it! What you're doing is changing the polar response and frequency response of the mic. For example, a vocal mic typically has a cardioid response; however, by cupping your hands around it, you push the normal cardioid shape back. The amount of sound the mic picks up is still the same, but what is being picked up is coming from farther behind it—and this is normally where the wedges and main PA are hanging causing feedback. Also, cupping your hands creates a resonant chamber, which acts like peaking EQ and causes feedback at the resonant frequency.

Sometimes when you are listening to your wedges, you might be able to hear the ring in the wedges, but can't hear it very well. In this case, you might actually intentionally change the polar pattern to cause feedback, but changing the polar pattern also can change the feedback frequency, so that method isn't really a good idea.

MIC PLACEMENT

Whether you're working with a brass instrument, drum, or vocals, there is always an optimum position for your mic. The question is how to find it. But it isn't as hard as you may think—all drums work on the same principle, and all speakers work on the same principle.

The goal, of course, is to capture the best source sound you can. Unfortunately, unlike in a studio, where you have plenty of time to play around with different mic positions, close micing is your only viable option in a live environment. Close micing reduces the amount of ambient noise and spill picked up by the mic. It is an art and is more than simply placing a mic close to the source sound. This is the third most important part of any sound chain, the first being the sound source and the second being the microphone. You may think that placing the mic right up against the speaker or drum would collect more sound, but the sound of an instrument takes some distance to develop, and by mic'ing too close you are actually missing parts of the instrument's spectrum and introducing the proximity effect.

When mixing a live show, the major factor you need to worry about is whether any mics are in the way of the performers. A live show can be a very exciting place to be, and if you have mics sitting over a drum where the drummer could clobber them, mic stands that have boom arms sticking out could hit the guitarist as he comes flying past, or cables that are draped across the stage floor could trip up the singer. None of these things really help your accurate mic placement because a tech will run back on stage and then stick it in front of something that doesn't need mic'ing up in the first place and all that hard work is wasted. As much as trying to get the right placement for the right sound is important, their practical location is actually more important.

FIGURE 13.4
Musical instruments are designed to sound good at a certain distance, where all the parts of the instrument contribute to the sound. Unfortunately, we aren't able to get the perfect position because we aren't in a studio, but moving the mic back just a handful of inches will make a massive difference in the tone quality picked up.

Here are a few tips to keep in mind when placing microphones:

- When mic'ing drums, you'll always get bleedthrough from other drums. Try not to worry about it—it can be controlled by close micing and gating.
- Make sure your vocal mic is behind the FOH speaker line. Look at the mic position, and then look at the speakers. From the mic position, if you can see the grill on the front of the speakers, try and move the mic back slightly.
- A mic picks up sound mostly from the direction it's pointed. So if a mic is pointed at the center of the piano, the mic may not pick up the whole piano. Use multiple mics and blend them together to create the overall sound you want.

Most of the time, you'll find that mic placement is part luck and part knowledge; it can be as much about trial and error as it is about accurately placing the same mic in the same place every day. There are so many variables that change on a daily basis that it can be nearly impossible to get the same sound every day, unless you're using regulated power supplies for all your amps: the same mic, the same cable, and so on. Knowing how microphones pick up sound, how speakers and drums give off sound, and making the two work together will give you the best result. Remember to be open to bizarre combinations of microphones and placement—you never know what might work best in any circumstance. (That said, if you're working with a corporate pop act, stick to the conventional rules. They don't tend to like a maverick.)

Now we'll go into some more specifics about microphone placement.

Microphone Axis

When placing a mic, the next thing to understand is its axis and how it responds to a signal. Picture a guitar amp where the speaker of that amp is pointing

FIGURE 13.5
This is a picture of a mic underneath a snare drum. As you can see from this angle, the mic body is at a 90 degree angle to the skin. The mic axis is pointing at the skin; the skin is on-axis to the mic.

straight at you. When you place a mic directly at a 90 degree angle, so that the source sound is pointing directly into the capsule of the mic, the source is on-axis to the mic. If you decide to turn the mic slightly, so that it points across the source sound rather than directly at it, the source is off-axis to the mic.

When a source is on-axis to a mic, you get a much fuller signal; the amplitude is at its highest, and the frequency response is at its best. Then, as the mic is angled away from this point, so that it's pointing more along the source sound, it starts to lose level and begins to pick up more of what the rest of the speaker and room are doing. Neither on- or off-axes are inherently better; the important part is how you apply them. You will also hear that mics have good or bad *off-axis rejection*. This is just in relation to the sounds being picked up from the sides of the mic. If a mic is picking up a lot more sound from the front than from either side, the mic has good off-axis rejection.

Attack Zone

Every type of percussive instrument has an attack zone, which is where the instrument is hit. On a drum, it is the center point of the top skin; on a piano, it is where the hammers hit the strings. The attack zone offers a wide range of dynamic behavior and is the part of the instrument where the players connect musically with their instruments. However, if you only mic up the attack zone, you'll only get that initial sound hit, and not a lot else. The entire instrument is part of the sound, and it's the blend between the attack and the tone that you're looking to emphasize.

FIGURE 13.6
This is a picture of a mic on a snare. As you can see by the position of the mic to the sound source, the snare is off-axis to the mic.

The Players

Before you spend ages seeking the perfect sound, you must realize that the guitarist, drummer, or pianist is the one who makes the sound. Everything down the signal path is affected by what they are doing. You can get your techs to play the instruments and get the rough sound, but you will probably have to make some adjustments when the musicians come out and play. Ultimately, you're at the mercy of the players—no matter how good you get something sounding, it's up to the artists on stage to really make it sound great.

MIC'ING DRUMS

There is no such thing as a bad-sounding kit, and with the right combination of drum tuning, mic placement, and EQ you can be well on your way toward making a drum kit sound better than it is. The real key is to get the correct tuning in the first place. Various different amounts of drum key turning, as well as different sizes and shapes of gaffer tape or other damping materials, are needed. Remember that the sound you hear when playing a drum kit is different from the sound being picked up from the mic—it might sound perfect when you're sitting there, but it could be a different story at the mixing console.

Old drum skins can be dull, so if you're having trouble getting definition in your drum sounds, check to see if the skins are old. If they are, replace them; new strings and skins are inherently a lot sharper than when they have been used. (Anyone who can play well will know this, and their style should be able to incorporate it.) Be aware of cheap drum skins—they can sound poor when first put on, kind of like a ping. Good quality skins, however, will tune up pretty well when first put on. Of course, they will need stretching, so make sure this is done well before soundcheck.

You aren't always going to be able to get the separation you need in your drum mics. If you have two drums that are tuned close together, you won't really be able to differentiate between the two. Tuning one of them up or down and giving your frequencies more space to breathe will help in achieving that extra bit of separation you want.

As far as sound goes, drums are fairly easy to understand. You have two parts of the total sound: attack, which is the initial hit from the stick, and tone, which is the ringing sound the drum makes. Usually you'll encounter drums with both the top and bottom heads on, but occasionally you might see the toms with only the top skin on and a mic placed inside it. This depends on what the drummer wants to do and gives you different options to work with. Mic inside captures more tone; mic outside captures more attack. The relationship between the top and bottom skins and the shell of the drum is extremely important: The top skin gives you the attack and the tone of the drum, and the bottom skin gives you the resonance. This is why you mic the top of the drum rather than the bottom.

If you have an excellent drummer, he'll hit roughly the same spot all the time. This is your attack zone, and you can identify it by the marks on the skin of the drum. If you mic just this part of the drum, though, you won't get much tone at all—the edges are where the tone is. As a result, the problem with close micing drums is that you are compromising a lot of elements of the drum sound. However, because you're physically restricted as to where you can place the mic, there aren't a lot of options.

Kick (Bass) Drums

In many rock and pop bands, you'll notice a hole in the front of the drum skin, usually below the band's logo. This isn't there just for decoration; it's there for a very sound reason: This is for the mic. Depending on the type of music, micing a drum without a hole results in less attack being picked up by the mic, which means there won't be enough definition between hits of the drum.

A very common occurrence is to use two mics in your kick drum. For more pop and rock bands, you may want to consider using two mics. The combination of the two mics can give you a far more versatile sound and also give you more attack, which some bands quite like. A combination of a dynamic and a condenser mic works well—an SM91 on the inside of the drum (make sure it's raised off the bottom of the shell on a cushion or foam to stop any handling noise when the kick drum is hit, and also to get a different mic position), and a Beta52 in the hole (both from Shure). Try not to put the diaphragm of the mic directly in line with the skin of the drum; the air pressure at the hole's edge can get pretty high. Even for a mic that's supposed to be able to handle high air pressure, it can cause popping. You can place your internal mic almost anywhere, close to the front skin or right toward the back. The attack from the drum is coming from the front head, so the closer the mic to that skin, the more attack you'll be getting—moving it farther back will cause you to have less. You should concentrate on the internal mic first, and then go and place the external mic. You can get phasing problems here, so the ability to be able to move that mic is important. Some engineers don't particularly like having two mics in the kick drum. But to be honest, it's about the sound rather than what is good or bad practice. Using one mic can give you better results, but it really does depend on what kind of sound you are after. Play around with different mics. There are loads of great ones from companies such as AKG, Audio Technica, Audix, Shure, EV, and Sennheiser.

Snare Drums

The snare drum is the main crack behind the drumbeat, and it's the low beat of the kick drum and the high beat of the snare that are so important in any drum sound. These are the fundamentals that make up the backbone of the rhythm section and keep everyone in time and on track.

The snare is actually the metal wires that are tensioned underneath the drum over the bottom skin, adding a rattle to the sound. This element of the snare drum sound is sometimes so often overlooked. Personally, I like a tight snare drum sound, one with a lot of body that sounds like a shotgun. The reason for this is that in a live environment subtleties can get lost, so if the drummer is a little shy in playing a few notes the sound can disappear behind other sounds and the drive can lose its push. At least by having a snare drum sound that tight it will always be cut through no matter how hard it is hit. To achieve this kind of sound, you need to have the snare under a lot of tension. More often than not a really loose snare drum can just sound bad; the sound is actually comparable to a biscuit tin. But there is a fine line between what sounds good and bad, and what works well in a studio doesn't always translate well to the live show. Snare drums are a very stylistic and personal thing; whenever you listen to different albums (even from the same band), you'll hear the massive difference in snare drum sounds within the album, and also across the same genre.

Mic'ing a snare can be tricky. All the drum hardware is in the way, and possibly a few mic stands, so it's always best to try and get this positioned when the kit is being set up. The top snare mic has to be at the edge to avoid being hit, but you can still achieve different sounds depending on where you point it. Pointing the mic toward the center, at quite a wide angle from the rim, lets you pick up a lot of what the drum is doing, including the attack from the top skin. As you pull the mic into a steeper angle, the attack is deemphasized, and you begin to hear more of the drum tone.

You can also mic up the bottom skin of snares. When positioning this mic, try using your hand to feel for the best place for the mic; this way you aren't going to get your head stuck in between the drum stands, or get your ears blown off by the person playing. The best time to do this is in rehearsals because you'll have the time to find a good spot and then somewhere you can aim for each day of the tour. So arm yourself with a set of earplugs and find that good spot. The air is pushed away from the bottom skin by the force of the drum being hit, and you will feel the air move most opposite the point of impact (which is usually where the drum stand is). Although it's probably a little impractical there, you have the point on the bottom skin you are trying to aim for.

I personally prefer a very steep angle on the top mic. I usually use something like a Shure Beta57, close to the rim, pointed just on the outside of the halfway point between the middle (attack zone) and the edge. This captures plenty of the sound and gives you both attack and tone. This angle, though, is sometimes very unusable because of the way the kit is set up. In these cases you have no other choice than to flatten the angle of the mic, making it more off-axis. Then I use an Audio Technica AE3000 on the underside, about 3–4 inches away from the sweet spot, which I find to be somewhere around the center. Remember to reverse the polarity of the bottom mic so that its signal adds constructively, not destructively, to the signal of the top mic.

Toms

The two techniques for mic'ing toms have everything to do with distance. With larger floor toms, you often want to get a bit more boom out of it; if attack isn't extremely important, you can achieve this by setting the mic close, on-axis to the edge of the skin. This will give you the proximity effect with some bigger bass frequencies and will also cut down on the attack from the center of the skin. The other technique, which is more common with pop artists, is placing the mic in a higher position, about 3–5 inches away from the skin, and pointing toward the center point of the drum head. This gives the sound envelope time to take shape and also gives you clarity on the attack. The only problem with having them this far away is that you can get a lot of bleedthrough from other drums.

A Sennheiser 421 on toms sounds absolutely stunning. Unfortunately, using several tom mics requires quite a few stands, and, being on the large side, these mics are difficult to get in position. The best position for them is directly over your toms. The next best thing I think is an EV 468, which has a supercardioid polar pattern and good off-axis rejection. It also has a tilting head, so you can easily get in and change the direction in which it's pointing (although I keep losing the screws for it). It also has a high SPL capability, rated to 144 dB SPL, making it ideal for loud drummers.

My general rule with mics and drums is not to use a mic with a small diaphragm. You are mic'ing up something that has a large surface area and the ability to move a lot of air. It's the movement of air that gives you power, so you want a diaphragm that's going to translate this to the PA, and large-diaphragm directional mics tend to have better low-frequency response than small-diaphragm mics.

FIGURE 13.7
An EV 468 on a Rack Tom.

Hi Hats, Overheads, Cymbals, and Percussion

Hi hats are the most percussive cymbals in a drum kit. A lot of musical time is kept on these cymbals, but sometimes they aren't captured as they should be. Mic'ing individual cymbals is very much like mic'ing a drum—you need to make sure the mics don't get hit and are pointing in the right direction without having too much sound from elsewhere. There are three common techniques, and the one you use completely depends on the drummer. If there's a lot of hi hat work and a lot of opening and closing of the hats, mic up pointing 90 degrees on-axis in the middle of both hats (Figure 13.8). This gives a percussive whoosh as the air rushes past the microphone and can add another deeper beat to what's being played on the hats. Just be sure that the mic does not pop. Another technique, which is employed by the majority of engineers, is to place the mic at 45 degrees above the hats (Figure 13.9), pointing either toward the bell of the center hat, or more toward the edge. The sound from the bell is more defined, and as you move toward the edge you get more cymbal hiss, like the classic hi hat sound. Changing the angle of the mic will increase or decrease the off-axis pickup. Changing this angle will give you a way of combining definition and cymbal hiss. The third technique is to mic from underneath. Usually this is done on-axis, but from underneath it gives you less attack than if it were placed on top. The advantage of micing from underneath is that you can get closer to the cymbal, cutting out more noise from the surroundings.

FIGURE 13.8
An audio technica AT4140 pointed 90 degree at a pair of hi hats.

FIGURE 13.9
The Classic 45 degree angle on the hi hats.

In the studio, you see producers and engineers using various cross mic techniques to capture the room and the drums. In a live environment, though, you just want to capture the cymbals, and maybe some drums. I have a pair of AT3035 (Figure 13.9), which are wonderful as a pair of overheads—not too sharp, but they also have good response in the low-frequency range. This is important because I EQ a lot out of my drums to give them punch and attack, and because I don't really EQ much out of my overheads (apart from the high-pass filter), they can add in some of the natural sounds of the drums and can soften them up a bit. This still leaves the definition in the punch and attack of the drums, but can make the drums seem like they are sitting back in the mix a little more than they actually are. Again, these are large diaphragm microphones; smaller mics are more directional, and with these I want to pick up a wide spread. Another mic to mention here is the AKG 414; it's always a good mic for these kinds of applications. Try flipping the polarity on both overheads at the same time; this will make sure they are sitting nicely with the rest of the kit.

As far as positioning goes, I like to have the overhead mics wide over the left and right parts of the kit, to give an extreme stereo image, while also keeping them just high enough over the cymbals. This also helps make your kit sound a lot bigger, when panned left and right on your console.

Ride cymbals are percussively similar to hi hats, and you can use a couple of techniques to capture what the drummer is doing. First, you can apply both top and bottom mic'ing positions, but pointed toward the bell if the drummer uses the bell a lot. However, it's often best to use the second technique—mic'ing underdeath—more often, as you might not have space to get the mic in the right position over the top.

Percussionists tend to use lots of different instruments—from bongos and congas to triangles and finger cymbals—and it's your job to make them all heard. The best course of action in this type of scenario is to mic up the static instrument individually, and then also give the percussionist another mic for hand instruments. If you don't have enough channels, you can even incorporate that microphone as an overhead, so it picks up a section of the percussion setup rather than individual pieces.

Remember that a figure 8 polar pattern has much less pickup from the sides, so if you're trying to mic up a hi hat and a cymbal (for example), it could be worth trying a figure 8 pattern mic rather than two cardioid mics to reduce spill.

Steel Drums

Steel drums (or steel pans, as they are also known) are fairly easy to mic. The sound comes from the pan that is being stuck, and the rest of the drum just gives the pan its tone. The bigger and longer the drum, the deeper the tone. The best course of action here is to use a small diaphragm condenser mic, positioning it underneath about 3 inches away from the bottom of the pan. This gives you all the attack, tones, and notes that are needed. A Shure SM81 is good for this.

MIC'ING SPEAKERS

Speakers change their tone from the center to the edge, although you have to be standing very close to hear it. Various factors should be considered in deciding where you should put your microphone. If you go more toward the center, you'll find there's more brightness or sharpness, whereas when moving toward

FIGURE 13.10
A pair of audio technica AT3035s as overheads.

the edge, you'll find that you start to get more of the sound of the tone of the cabinet. The guitarist has normally chosen the amp and cab for its sound, so you should be aware of this when mic'ing up.

When placing a mic in front of a speaker, you need to figure out at what distance you want the mic to be placed. In live environments, you usually want your mic anywhere between right on the grill of the cabinet, up to about 6–7 inches away. You'll hear the difference as you move the mic backwards. Having the mic closer to the cab causes it to have more bottom end, due to the proximity effect; as you move it back, the wavefronts have a little more distance to form, so the tone starts to change. If you have a harsh guitar sound, moving the mic closer to the cone edge can help because that reduces the pickup of highs; If you are having trouble seeing the speaker through the grill, grab a torch and shine it into the grill. You'll then see the speaker light up behind it, helping you to do your mic placement.

When you're dealing with speaker cabinets that have more than one speaker, you need to make sure you're concentrating on one speaker at a time. Don't start sticking the mic in the middle thinking you're going to be picking up sound from all of the speakers—you'll only be picking up the sound of vibrating wood.

If you want, you can double mic the guitar cabs. This decision depends on how important your guitar sound is to the mix. If you're going to do this, you really want to put two different types of mic on the cab—try a combination of a dynamic and a large diaphragm condenser mic. You'll get the smooth gritty tones of the dynamic, coupled with the sparkle from the condenser. Be aware of phase issues, though; you may need to pop in the polarity switch. To check for phase problems, pull one of the faders down on your console. If the sound suddenly jumps to life, you have a phasing problem. Pop in the polarity switch, and push the other fader back up again—this should theoretically give you one kick ass guitar sound. If it doesn't, go back to the mics and try moving one of them slightly backwards or forwards. This is a little hard to do without a point of reference, so use a wedge or a set on IEMs, and get the monitor engineer to send you both mics. This applies to any sound: the more mics you put on a sound source, the more likely you are going to run into phase problems.

If you have an indie-type band that likes playing through open-back guitar amps, you can try micing up the front and the back of the speaker. You'll need to flip the polarity on the back mic as well, and experiment with different positions.

Be aware of mic'ing too close to the floor, which can result in bounceback and bizarre responses in the low and midrange (though this does depend on the type of floor and how it's finished). If you're having trouble getting rid of a frequency you don't like, try using a different speaker. Alternatively, if it's an amp and a speaker combo, angle it slightly up, away from the floor, or move it higher up.

These techniques don't just apply to mic'ing guitar amps; they also apply to anything with a speaker in it that's pointing toward you, such as a bass or keyboard amps.

MIC'ING ACOUSTIC GUITARS

Sometimes, out of the blue, one of your artists will walk into the gig and present you with an acoustic guitar. They probably spent the previous night playing a really old, beautiful-sounding acoustic guitar that was made by a man in a shed in Dorset using nothing but matchstick boxes and chicken wire, and now they think it's time to make the set a little bit more serious. The only problem is there is no pickup, and you are going to have to mic the guitar.

You can get two types of strings on an acoustic: nylon and steel. Both bring a very different sound to the guitar; nylon string guitars don't have a pickup inside them, as nylon can't disrupt the magnetic field around the pickup. Occasionally these types of guitars might have mics inside them, but they aren't always of the best quality. Steel string guitars are used a lot more and generally lend themselves to being used live a bit better than nylon stings (but it does depend on the type of music).

Along with the type of string that makes up the sound of an acoustic, three other elements make up the whole sound: the neck, the sound hole, and the body. They all have their own influence on the sound of the guitar, so you aren't going to get the best sound you can by using just one mic. Micing the neck will give you a lot of the higher tones, whereas the mid tones come from the body, and the full-sounding body of the guitar comes from the sound hole. In addition, the attack zone is where the guitarist strums, using either a plectrum or fingers (which give very different sounds). Plectrums (picks) result in more attack than fingers; it gives a sharper sound, and the initial notes cut through more. Fingers, on the other hand, are a lot more subtle; they offer a much wider range of dynamics and can sound more intimate.

Usually, an acoustic guitar in a live setup will have a pickup, so you can simply pop a DI down and plug it straight into the PA, saving yourself a lot of bother. Depending on how influential and important the acoustic guitar sound is in the whole mix, you might consider doing a combination of DI and mic. If you're going to do this, use the DI to get the fullness and look after the lower frequencies, and concentrate a small diaphragm condenser mic at the bottom of the neck to pick the higher tones off the strings. This mic should be placed around 6 inches away from the guitar, so as not to interfere with what the guitarist is doing.

If you don't have a pickup, use a small diaphragm mic at the neck position we just spoke about. However, you will probably have trouble getting any thickness and warmth in the sound so you could try moving it toward the sound hole, but chances are there is going to be a hand in the way. In this case, if you can use two mics, try using a larger diaphragm condenser on the body. Avoid using anything other than a cardioid pattern; otherwise, you could end up with all sorts of feedback issues while trying to warm up the sound.

With any guitar, be aware of old strings, just like old drum skins; new guitar strings can sound so cool (this goes for any stringed instrument). Another handy little thing to have in your acoustic guitar case is a feedback buster. It's made of rubber and covers the sound hole in the guitar. This only works with acoustics that have a pickup or mic inside because it cuts down the interaction between the sound outside and inside the acoustic, which means you hopefully get more of the level you need. (That being said, any decent pickup in a good quality acoustic shouldn't have this problem.)

MIC'ING VOCALS

The singers' mic technique is always a love/hate relationship. Just when you think they've got it right, they go and mess it up again. It is your responsibility to show these so-called singers that they can't wield their microphone like it's a bottle of vodka. After all, they are ruining your mix!

When mic'ing vocals, you want to smooth out any lumps in the sound so as not to have too much low when they pull the mic close and when the mic is pulled away the sound doesn't become tinny. You also want to stop any distortion that might occur when the singer belts out a note too close to the mic and stop blurry bottom end from occurring when the singer is singing weakly in a low range with lips pressed hard against the mic. Don't worry too much about getting a level sound at this stage; just make sure the singer understands the technique first, and then the sound level will quickly follow.

Get your singers to listen to how their voice sounds with their lips on the mic, and then pull it away and make them listen to the tinny sound that comes from the speakers. This seriously works; most singers have an ego, and they all think they sound great. By pulling the mic away too far, they start to sound like a robot from a 1950s B movie. This is not what they want as a vocal sound. Once this has sunk in, tell them that by using small movements to and from their mouth, they can emphasize important parts in their songs. The development of the mic technique is very important not just for them or yourself but mainly for the audience.

In my opinion, dynamic mics work best for vocals. Although condenser mics are very good, high-quality mics and are actually better for preserving the natural tone of the singer's voice. A dynamic mic can be much more sympathetic, more forgiving to a bad mic technique, and also tends to have far better gain before feedback, not picking up anywhere near as much ambient noise and spill than condenser mics.

MIC'ING PIANOS

As with most types of instruments, there are many different types of pianos, all offering very different sounds. Grand pianos are frequently used in concerts. One common debate is whether to keep the top of the piano open; you can get a sound that is more resistant to feedback with the lid closed, but it can also be a little boxy. Personally, I prefer keeping the lid open and then trying to get the optimum mic position. You might also be presented with an upright piano, where all you have is the soundboard in the back to mic up.

The problem with using pianos in a rock show is that you have to mic an acoustic instrument in the midst of amplified ones, which can lead to bleedthrough into the piano mics. However, you only get bleedthrough if an instrument is either louder or quieter than the rest, so make sure the stage level isn't ridiculously loud to give yourself a fighting chance.

When you look at any acoustic piano, you can see the string running across the length of the instrument—starting with the low register strings one end and working up to the higher register strings. You'll also notice that the hammers that hit the strings are usually situated behind the keys. In the case of a grand piano the keys are at one end of the strings, whereas in an upright the keys are in the middle of the strings, and this is where the hammers hit. The point at which the hammer hits the string is your attack zone. (With a piano, though, attack is a lot subtler than with other instruments.)

A piano has an enormous range of dynamics, from large crescendos to sharp staccatos to soft dal nientes. Coupled with this range is an enormous frequency range. Due to the nature of the directionality of mics, and the area that has to be mic'ed up, two mics usually aren't enough. (Engineers sometimes use a crossed pair configuration, but this picks up ambient noise.) To be honest, I'm not sure there's a right way to mic up a piano at a show. Most of the time it's all trial and error, and what works one night might not work the following night because of the ever-changing acoustic environments. But a combination I have found that works quite well for live environments is a five-mic array. Depending on the size of the piano, I'll use three small diaphragm condenser mics about 5–6 inches away from the attack zone to capture as much of the defined notes as possible, then two large diaphragm condenser mics to capture the sound of the body. Remember, the farther you get from the attack zone, the more body you get. In the case of a grand piano, having two mics just over halfway down the piano and about 7–10 inches away should help you capture some nice full body.

Pianos are tricky. Sometimes the pianist will play at such a low level that you can hear room noise if you turn the mics up to compensate. In addition, sometimes it's hard to control the low end, and you have to roll in your high-pass filter and lose some of the guts of the piano. If you're having problems getting certain elements of the piano out, and different mic placement isn't helping, think about inserting pickups in the problem places in the body. This way you can use a pickup–mic combination to get what you need. (Some people might think this is a little sacrilegious, but it's better to be able to hear what's going on than not.) If anything, it's the acoustics of the room that will cause problems. If you look at well-built classical concert venues, you'll see that they have a lot of acoustic treatment, making it easier to mic up these kinds of instruments.

MIC'ING BRASS SECTIONS

The always-popular brass section (horns) brings another dynamic to music. Mic'ing them is fairly easy; obviously, the sound comes from the bell, so this is what you want to mic. The sound on-axis to the bell is brighter (edgier) than off-axis to the bell.

Because brass instruments are held, they're going to move around, so the question is: Do you let the musicians control their dynamics, or do you want to? If you want to let them do it, you have a whole range of mics at your disposal: the Audio Technica Pro25, the Sennheiser 421, the EV RE20, and so forth. These are excellent mics that bring out many different qualities in sound. These all go into a mic stand, so the musician controls how close or far he is from the mic. I personally prefer this approach because it lets brass players get a feel for the music, but there is a mic technique that they need to learn as well, which is pretty much the same as the singer's mic technique, but nowhere near as subtle. It's all about emphasis; you see, when they are playing a solo, they automatically push their instrument forward. If you can get a mic position that works for you at their natural rest state, then when any emphasis for individual notes or a section of a song comes along, they'll move forward and back giving you some nice dynamics. The only problem with this is that when a brass instrument is played, the hearing of the player is closed down so the sound they hear is the resonance of the instrument in their head.

The other approach is to use a clip-on mic, which attaches to the edge of the bell of the horn, and then is bent around to face into the middle part of the bell. This keeps the mic at the same distance at all times, so you're the one who controls the volume, and the player's feel comes from his breathing technique. This technique can, however, pick up lots of tapping as they are playing away. Good quality shock-mounted mics should be able to isolate the mic from the body of the instrument, though, so this doesn't occur.

DIRECT INJECTION (DI)

The DI box converts all your line signals into balanced mic line signals. There should be a PAD on each one, which reduces the level going into the DI box if the input is a little on the heavy side. It also has an earth lift function, which can eliminate any hums or buzzes created by ground loops. DIs come in two types, passive and active. The main difference between the two is that passive DIs don't require any power, whereas active DIs do. The difference is like that between dynamic and condenser mics, and both types offer two major advantages on their own. Passive DIs have the advantage of providing extra isolation from the input to the output because of the transformer-based circuit they use inside them. This means that they can be more efficient in reducing any hums and buzzes on the line. Active DIs require power to work; this can come in the form of phantom power or a battery. The advantage of using an active DI is that high-end frequency loss is minimal compared to using a passive DI. Because the input in an active DI is very high impedance, it does not load down the guitar cord and pickup. The lower input impedance of a passive DI can roll off the highs and/or lows.

In PA terms, bass guitars are less reliant on using sounds from an amp; at a live show, it can prove to be rather difficult to get the serious bottom end you need from the bass amp's speakers. As a result, most channel lists have a bass DI and a bass mic. Obviously, the bass player will have his own setting on the amp, which you'll want to capture, and when the DI and mic are used in conjunction with each other and blended, they make one full and deep bass sound. You'll get a lot more presence from the bottom end, which is also a lot more controllable.

Keyboards are always best DI'ed. Leave the keyboardist to set his or her own level on stage if they have an amp, but take a direct signal. Because these instruments cover such a wide frequency range, it's important that you don't lose any frequencies not being reproduced by speakers.

I once toured with a band whose guitars sounded terrible because they didn't have the right pedals to use their amp distortion, and the band also insisted on using pedal distortion (and this was the fundamental problem). I played around with the sound on the pedals for ages, but just couldn't get them right; what would happen is midway through a song, they'd kick in their distortion, and all of the dynamics would be sucked out and replaced by what sounded like a wasp trapped in a jar. Then, once the distortion was turned off, all the dynamics would come back. I could get it to sound heavy, but I couldn't make it intelligible. The problem with a lot of cheap distortion pedals is that they process the sound in such a way that the original signal is squashed. This causes the extremes of the frequency range to disappear, and a kind of compression occurs so the natural dynamics of the instrument are squashed as well. To try and fix it, I decided to add a DI box before the distortion pedal, and then bleed a little clean guitar into the heavily distorted parts, just to give the sound a little bit of clarity.

Bassists sometimes use pedals as well. The fundamental problem with all types of pedals is that they squash the frequency spectrum, which doesn't always show up in the case of guitars, but when dealing with bass guitars it does. When you stamp on your distortion, it might sound great through the amp, but you also lose the extreme low end that was driving the mix. By trying to push the sound over the edge and making it huge, you're actually grabbing hold of it and pulling it back to safety, and that's not very rock 'n' roll. Always use amp distortion where you can; if you can't, think about taking a DI line. In the case of bass guitars, make sure you always have the DI line before the effects; then let the mic on the amp or another DI line pick up all the extra effects.

IF ALL ELSE FAILS...

Mic techniques are a very personal thing. A technique that works for one person won't always work for someone else. You should always try a few different combinations of mic and position just to hear how the sound changes. Sharpen your listening skills; above all else trust your ears.

The most important part of any audio chain is the beginning. Get that right and the rest of the sound will fall into place no matter how good or bad the

FIGURE 13.11
If all else fails and you are having trouble picking up a good sound, tick the biggest mic you can find on the source.

equipment is. There are some amazing mics out there that will help you do a fantastic job, from companies such as Earthworks, Sennheiser, Audio Technica, Audix, and EV, so try as many mics as you can on as many different applications. You never know what you'll hear.

From time to time, especially in the early days of your career, you're going to be presented with a mic case that is full of beaten-up mics of all shapes and sizes, and sometimes consisting of brands you've never even heard of. If you're lucky, there will be at least a couple of Shure SM58s or SM57s in there; these are industry standards for a good reason. They're cheap, you can put them on anything, and they do the job. You might not get the best sound you are looking for, but these will help you get out of any situation, so use them if you need to.

THE PATCH

Once you have all your mics on stands and in position, the last thing you need to do is plug them all into the PA. You'll notice that the channel list you made earlier is numbered in channels and that each channel has an instrument dedicated to it. The mic cables run from each mic and must run into the corresponding channel on the mixing consoles. This is known as the *patch* and is the physical plugging in of the mic cable.

It's really important to do this properly; otherwise you could spend ages hunting around for a mic that is plugged into the wrong channel or not plugged in at all. The problem can come from when you're using satellite boxes run from the other side of stage that are then plugged into your stage box. Let me explain: Your guitar is on the far side of stage and should come up on channel 12 on your mixer. It has been run into line 5 on your satellite box, which means that line 5 from your satellite box needs to be plugged into channel 12 on your stage box so that it comes up in the right channel on the mixer. That might sound like a fairly simple procedure, but when you are faced with multiple lines from all over the stage going into different channels, it can get very complicated.

At festivals, you might be given what is known as a *festival patch* for the consoles. After you send your channel list, it will be passed on to the company supplying the PA, and they will then integrate your channel list into their master channel list. This means that your channels are going to be spread out so that they incorporate all the other channels' list of the bands that are playing the same stage, and may no longer run in the same order as it did before. This does actually save time during change overs and may make it easier for the patcher to get the patch right for each band because all the channels will be grouped together, and on your console you'll have your dynamic processors plugged in at roughly the right place.

CHAPTER 14

Soundcheck

In this chapter I'm going to talk about a soundcheck and the processes that lead up to it. By this point your PA and mixing console should be up, all your mics should be in position and plugged in, and all your DIs should be present and correct.

COMMUNICATIONS

Before any soundcheck, you must make sure you have good communication with the stage. There is nothing more frustrating than having to shout, or run up to the stage to try and make yourself heard, and it's pretty hard for the guys on stage if they can't communicate with you. You want to set up your *talk to stage* (TTS), or talk-back as it is otherwise known. This is set up using a mic at FOH, which is sent to the monitors so that everyone on stage can hear you. The TTS mic signal will either be sent to the monitor console (if you have one), who will then feed the monitors, or your aux sends (if they are feeding the monitors).

A *com system*, which consists of a series of headsets with microphones attached, is always a good idea for personal communications between FOH and the monitor world. This way no one will be disturbed when you are trying to sort out problems (which could be a rather lengthy conversation).

We also sometimes use what is known as a *shout system*. This is a very popular option with festivals because it is the fastest way to communicate between FOH and the monitor world, and is used in conjunction with the previous systems as well. There are speakers and mics at both FOH and monitors that are connected to each other, and they are left open at all times, meaning that you are able to talk into a mic and the sound comes out the other end without having to ask anyone to turn it on. It is normally used for asking questions and for asking the various engineers gathered to go to coms for a longer conversation.

Then there are two-way radios. The whole production and not just the audio department usually use these. Most of the techs I know loathe the day they are given a radio; it means the lazy production manager doesn't have to leave his desk to come and find you, as you are now only a click away. But they are very handy during a soundcheck when you are trying to communicate with the monitor engineer, the tour manager, and the backline techs, all at the same time. I've been using radios a lot this summer during festival season because it wasn't always possible to get good communications between all of us using the other methods.

THE TAP ROUND

The first thing we want to do before we start making any noise is a *technical line check*, which is affectionately known as a *tap round*. This is the process of testing to make sure that all the lines are working and coming up in the right place on your console. This process is fairly simple and straightforward; all you need to do is tap each microphone in turn, for any DIs that are used, plug a mic into them (using an XLR to jack cable), and speak into or tap the mic. Start at channel 1, and then work your way through each channel in turn. Good communication is important here, so everyone knows which line they should be looking at. If you encounter any problems (such as buzzes) on a line, this is the time to sort them out. A technical line check is especially important if you have a lot of stereo channels, as you won't always be able to tell if the stereo image is correct.

LINE CHECK

After you've completed your tap round and confirmed that all the lines, mics, and DIs are working properly, you need to do a line check with the backline techs. Again, communication is key here. Make sure everyone can hear each other and knows what is happening and in what order. This is when the instruments are actually played and any changes that need to be made to the sounds of the instruments on stage, any mic placement changes, and any phase issues are sorted out.

When you are running through each instrument in turn, aim to get each channel's gain structure in the right ball park. Gain structure is the second most important part of your audio path after the sound source.

Your gain structure is about finding the optimum level between the floor noise (the inherent hiss in electronics) and the point at which the signal distorts and begins to fall apart. The important part of the entire process is that you are easily able to follow your gain through from your preamp to your amp. The trick with getting this right is to make sure that the master section (left, right, mono, auxs, subgroups, VCA faders) is all set at 0 dB, that anything in line or inserted (such as the graphic EQ and crossover) is set at 0 dB. This way, when you come to start setting the gain on the preamp, you can easily follow through the entire signal path on the console.

You may occasionally need to change to level on the master faders, but only if your output is too loud. As long as you have set your gain up properly on the channel strip, you should not get any distortion on the output stage. Having the right channel strip gain also helps when sending any signal out of the auxs.

Let's now have a look at how we set up the right gain structure on our preamps.

There are two pretty standard ways of setting your gain structure. The first involves your setting the level on the meter to 0 dB (unity) using the gain control knob. The other is that you set all your faders to 0 dB and then set your gain so that all the channels sound like they are at the same level.

The truth is that it all depends on how much you want to mix. If you have all your signals coming into the console at 0 dB on the meter, then you'll probably have a lot of the faders pulled down, which means the closer they are to the bottom the less control you have over the signal being too loud or too quiet, or you'll have your master faders pulled right down. If you push your fader to 0 dB and then adjust your gain from there, however, you'll have far better level control when mixing on the faders.

If the signal coming into the console is immediately going straight off the end of the meter and you haven't turned the gain control up yet, then you are going to have pop in the PAD button to reduce the level so that it is more easily controllable and won't be distorting.

The best way to set your gain structure is by using your ears. 0 dB sounds different on every console, and it's all to do with the way the preamp works. On some consoles the closer you get to the clipping point, the edgier and warmer the sound becomes, and on instruments like guitars and drums it can be very nice, but when you want a nice, smooth bass DI sound, it might not work that well. On other consoles when you start to breach 0 dB the signal starts to become harsh and distort fairly quickly and is generally unpleasant to listen to.

In the case of most digital type consoles, the signal can't distort because a digital clip sounds horrendous. Any digital console worth its weight in gold will have digital clip way above the point at which the preamp clips in their A/D converters.

Another thing to note before we move on is the word *headroom*. We use this word all the time, and it means the amount of level that is left before clip.

After you have gotten a good gain structure, do some EQing, and get a good solid starting point for the band to start their soundcheck. But remember that the backline techs don't play exactly like the band, so you'll probably have to do a bit of tweaking when the band comes out. While we are on that subject, it is also a good idea to point out that, in fact, the band usually plays differently between soundcheck and show. They will be more hyped up, and the adrenaline will be flowing at stage time, but there really isn't anything to get excited about at soundcheck.

Make sure to check any level differences produced by different settings on each instrument. If a guitarist uses different effects, make sure you run through each one, that each sound works with the EQ you have applied, and that the output levels on each pedal work with what they are used for. In rock shows, you may want to emphasize parts of songs more than others; using a distortion pedal with a slight jump in volume can add a loudness effect so that the audio appears to be bigger. Little jumps in things like choruses and solos help add that extra dynamic to a mix. If there is no jump, and it's all the same level, that's fine too; you can boost on the fader as you need to. However, when there is a drop in level, you need to worry. Reducing the volume also changes the sound; it becomes weaker, sinking the sound into the background, and your chorus or solo has a rather large anticlimax.

Your job during the line check isn't to balance the instruments and other audio; it's just to make sure that everything is present and correct, and that the tonal balance of sound within each instrument works as best as it can. This is your last chance to fix any technical problems. While you're doing it, a word of advice: Don't get the band out before you have to. Having delays while the band is on stage can lead to a very difficult soundcheck—everything should be running smoothly and calmly. If anything does go wrong, calm communication is the best way to deal with the band and the problem. It's important that a band has confidence in their gear and their crew; they need to have that confidence when they walk out on stage later, to give you the performance that you need to make them sound good.

STRUCTURING AND SOUNDCHECKING

Soundchecks are vital to the success of any show. Once you've got the correct gain structure and have done some EQing at the line check, you're ready to get your balance right with the band.

To begin, it's a good idea to structure the sound check; this is very important for the sanity of everyone involved and helps make sure you don't keep anyone longer than necessary. (The last thing you want is for the stage to be inhabited by a bunch of musicians just wanting to play "Africa" by Toto.) If you don't need the whole band at first, just get the musicians you need. It isn't always necessary to soundcheck everything individually. Some engineers like to check the sound between the drums and bass first, so get them to do a few riffs, followed by the guitarist, and so on. This will be especially helpful when starting out because you'll only have a couple of things to concentrate on.

You should also have an idea of which songs you need the band to play. Begin with slow songs; these are good in a lively room because the long notes are easy to define, and thus are easy to tighten up. Then move on to hard and fast songs, so that you can hear how the room responds to the whole mix. Make sure you listen to any unusual songs as well; perhaps, for example, the band plays one acoustic song. Essentially, you want to hear everything through the PA in song format, at least once.

FULL VS. EMPTY ROOMS

A very common oversight is the understanding of how a room behaves when empty or full. The differences can sometimes be obvious and other times extremely minor. You will be able to get some idea of how different it might be by just looking at space. The basic idea is, if the space is a large untreated concrete box with a lot of distance between the audience and the ceiling, it's going to be very difficult to mix in because there is still a lot of wall space for the sound to be reflected off. This means lots of natural room reflections clouding your mix, which aren't going to tighten up that much after the soundcheck. The Big Top in Sydney's Luna Park is one such nasty place. It is pretty much like an aircraft hanger: massive concrete floors and walls, with a huge distance between floor and the silver air conditioning ducts attached to the ceiling. If you think that the tallest people in the audience were just over 6 feet tall, consider that there was another 40 feet to the ceiling. This gives the audio a massive amount of wall and ceiling space to be reflected off, and although the room has been full of people, the reflections lost from the floor aren't going to make that much difference to the overall sound quality. The bottom end is still going to be swirling round, and you are still going to have a tough time getting the clarity out of your melodies. But with every challenge you encounter, the more experience you will get.

When you compare that space to a room with a low ceiling that has ceiling tiles and a polished hardwood floor, the difference is amazing. Most of the sound will be reflected off the floor because the ceiling tiles will absorb a lot of those reflections. Because the floor is polished hardwood, you'll probably find that there is an abundance of high midfrequencies. If you pull these frequencies out so that the EQ in the room sounds right for soundcheck, when the audience enters the room later and fills up the floor these frequencies won't be in the EQ at all and you are going to need to put them back in.

There is not only a difference between full and empty rooms but also between an empty room and one that is only half full. Always try and find out how many people you think are going to be at the show that evening, and find out how hot and humid it tends to get in the venue. We'll look into the heat and humidity of a show in Chapter 16. Anticipating how many people will be at the show will help you understand how the room will be changing. It's always a good idea to speak to the in-house engineers as to what happens in the room, and how the sound changes as the venue starts to fill up.

It is pretty impossible to tell where certain frequencies are coming from, but with a little understanding of how reflections work and how room acoustics change when an audience arrives, you'll be in a good position to start the show. Sometimes you might need to pull a few frequencies out of the PA just to get through a soundcheck, with the full knowledge that you'll be putting these frequencies back in once you have a floor full of audience members.

MONITORS

The most important part of a soundcheck is getting the right monitor levels. A lot of the confidence for that night's performance is built up during the soundcheck, and the band knowing that they can walk out on stage later and have a near perfect sound will give them the confidence to give a great performance from the word go. And it's the performance that we are after.

It might be a good idea at times (if the band is struggling with monitor levels) to turn off the FOH. In lively venues, this makes it easier to hear what's going on in the wedges. If a band is having serious problems, you might want to get on stage and listen for yourself. Sometimes it just takes a fresh pair of ears to hear what's going on. In these situations, you can find everyone getting a little frustrated, and as always, communication is key.

Now that digital consoles are getting cheaper and smaller, more bands are taking their own monitor consoles on tour. When touring with your own console, especially a digital one, remember that little changes on the console can gradually turn into big changes. Every once in a while, it's good to start over—or at least load a show from a few days ago. Rooms change, and these do affect the monitors, even in-ear monitors, which means you are adding slight variations in rooms on top of each other. Be mindful of the changes to the mix you are making from day to day.

INTERACTION BETWEEN THE STAGE AND FOH

A very strong relationship exists between the front of house and monitors, and getting it right is important. One of the fundamental factors of any good front of house mix is getting the stage volume to a decent level. Don't let the stage level creep up, as this can cause enormous problems with the mix at FOH. It can cause phase problems, resulting in weak kick drums or vocals, or making it so that nothing happens when you try to pull the bass down in a mix. The most obvious problem, though, is having the guitar amps too loud. Remember: Isolation of things like the guitar amps and drums is important; if you can improve it, you are well on your way to a well-defined and clear front of house mix.

Unfortunately, turning down the volume on the stage has its own problems. Consider the guitar amp situation: Valve amps have a valve in them, and at lower volumes they don't really do much. When you turn the amp up, though, the sound becomes warmer and warmer. To address this problem, you can turn the amps around so that they are pointing either across the stage or to the back of the stage; that way, the audience isn't getting any of the direct sound from the amps. Alternatively, you can get a Power Soak; this device sits between the amp and the speaker, and soaks up the power from the amp while still keeping a nice warm sound.

There is one more point about monitors that I should really go into, and this in a way is more important than getting the right separation. The way that the monitors are EQ'ed is really important in getting the right sound for the audience. Think about this; your FOH is EQ'ed so that the music sounds great through it, and your wedges may be EQ'ed so that a specific instrument sounds good through it (vocals, guitars, that sort of thing) Let's, for argument's sake, say 400 Hz is too prominent at FOH and is pulled out on the graphic, but it doesn't disappear; it only becomes distant. While 400 Hz might not be very prominent on stage, it might just be that some frequencies on stage (fundamental, harmonic, or subharmonic) are reacting with the main auditorium causing 400 Hz to be somewhat unEQable at FOH.

Frequencies interact with a room in all sorts of ways, from summing in certain parts to cancellation in other parts. Until now, we've only talked about it from a FOH point of view, but the monitors also affect what we hear in a room. If you're having problems getting rid of frequencies, turn FOH off; if you can still hear that frequency, it's likely that the monitors or a guitar cab spewing its noise right at your head is the culprit. Take a walk around the front of the stage and listen to how the audio is coming off the stage. You might find a point where that frequency starts to sum or cancel, which will give you a better idea of where it might be coming from. Get up on stage and have a listen. Re-EQ the monitors slightly if you need to, but you might just find that it is the side-fills, or maybe just one of the wedges.

GET UP, GET OUT, GET LOOSE...

While the soundcheck is happening and you're building your mix, you need to get out from behind the console. As we discussed earlier, the sound is different all over the room, so you need to have a listen from the audience's point of view. Do your best to make the sound as good as you can for as many people as you can. The hardest part of this is when you have a bad mix position—you will always be tempted to mix it for your ears rather than the people at the show, but don't make this mistake.

Getting out from behind the desk will help you hear how the sound is reacting in other parts of the room and will help make sure your mix is going in the right direction. Always check how it sounds in the middle of the room, which is where summing is most likely to occur and where you are most likely to get a flat, two-dimensional image that can sound very uninteresting because mixing when you aren't on center can cause you to compensate slightly for the off-balance of the PA. This causes the stereo image created by your EQing and panning to be slightly off center.

NOISE POLICE

Before we wrap up our soundcheck, we need to have a look at noise levels, as chances are you'll have the noise police imposed on you at some point.

The ongoing debate about sound limits will probably never end. With more and more government regulations all over the world, you will encounter more limitations on actual volume in a venue. We have official noise limits for places of work, and anyone exposed to high levels of noise is going to be given some sort of ear defender. We even have noise limits for protecting neighbors against hearing each other. Why in the music industry do we use industrial limits to protect people from something they want to go and listen to, whereas in clubs it seems to be something that is overlooked in certain places? Live music is supposed to be very enjoyable, and if you go and see a live show, chances are you actually want to be there. I do, however, understand that you must protect the people who are working in these environments, and when you are making lots of noise in a field, the elderly neighbors might not like having their afternoon tea falling off the table because of the vibrations caused by a youth wielding a guitar!

These noise limits don't have to destroy your mix, though. As long as you know what the rules are, you should easily be able to get a great mix. It's too easy to turn the whole mix up and make it sound powerful, but the ability to make a quiet mix sound great while still managing to create the space and the punch is the sign of a talented engineer.

Let me show you an example of noise regulations posing a problem. The Glastonbury Festival of contemporary performing arts is held once a year in a small town just outside the British city of Bristol. The main stage is at the bottom of a large bank facing up it, kind of like a natural amphitheater. On the other side of that bank are a couple of fields, and then the villages of Pilton and Glastonbury. The festival is renowned for its rain, and this particular year was no different. The headliner was on stage as the rain came down, and up came the umbrellas. Unfortunately, umbrellas with rain on them are a shiny surface, so as soon as the sound hit them, the audio went bounding up and over the bank, and into town. The locals didn't like this at all, and the noise police had a field day with their dB meters. The outcome was a reduced noise limit on the main stage, to the point where it was possible to hear the sound from the second stage coming from the next field. The authorities nearly had a riot on their hands because the audience couldn't hear anything. Eventually this was realized and the limit was lifted, but the situation could have been a lot worse.

The noise police typically turn up at FOH with their own decibel meter and set it up. The majority of readings are done from here, and this will be where you are told how loud the show can be. Readings may also be taken from outside the venue in order to protect the unsuspecting public. If you're doing some kind outdoor event, you might even have someone monitoring the sound level over the entire site; when one of the stages goes over the set limit, they bark commands down a radio to that stage telling them to turn down. A record might also be taken of sound levels throughout the show, so if someone decides to complain to the authorities, they have readings to look at. If the levels are within the stated boundaries, there's no problem; but otherwise you could receive a serious fine.

dB Meters

You measure sound level on an SPL meter. As we discussed earlier, we use the dB SPL scale to measure volume. 0 dB is no sound at all, and anything over 110 dB will have your ears ringing for days. SPL meters are supposed to have a calibrated microphone on them. (Using the latest iPhone app is not good enough, but can give you a very rough idea of what you are working with.)

I'm Weighting . . .

On every decent SPL meter there are at least two different ways of measuring frequencies; this is known as *weighting* and comes in *dBA* and *dBC*. These are two different types of frequency curve or filter. The A-type weighting is the most commonly used. It emphasizes frequencies between 1 kHz and 4 kHz, which is where you get all audible clarity, and is the scale most frequently used to measure hearing risk. A normal limit to see would be set around 95–105 dBA.

C-type weighting is mainly used to measure noise pollution. It hasn't been very common, but it is starting to be used more and more these days. It responds to pretty much the entire frequency band, just dropping off at the extremities. It is quite normal for you to have what looks like high dB limit in C-type weighting—say around 115 dB—though it actually depends on how the mix is balanced. Because nearly the entire frequency band is being measured, low-end frequencies suffer in this weighting. Generally, live audiences like more bass in a mix, but the more bass level you have, the closer to your dB SPL limit you will get. This can sometimes cause a few issues for engineers.

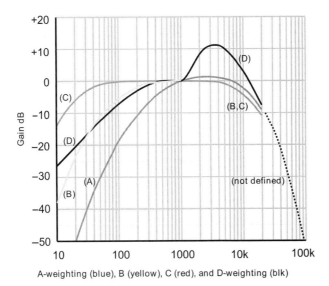

A-weighting (blue), B (yellow), C (red), and D-weighting (blk)

FIGURE 14.1
This image shows the dBA and dBC frequency curves we have just been talking about. Also, here you will notice the dBB and dBD curves. It is extremely rare that you will ever come across this, but if you ever do you'll know what the frequency response is. To see the color version of this image, visit the companion site at www.liveaudiobook.com.

Fast and Slow Response

The other function you will find on your SPL meter is the Fast and Slow button. This function measures how quickly the meter responds to the peak sounds. If you have a fast reading, you'll see the numbers changing very quickly on the display, whereas with a slow reading, the numbers change much more slowly.

Honestly, that's really all you need to know about the actual physical meter. They all come in different shapes and sizes, but those functions will always be the same.

Peak and LEQ

Following from the fast and slow response, we move on to slightly more advanced SPL meters that you will come into contact with at quite a few bigger shows. We use the dB SPL measurement and the sound's duration to tell you how damaging to hearing a sound is. The measurement of time comes in two forms: peak and LEQ. The peak reading tells us the loudest point we can go to, whereas LEQ takes an average reading over a length of time. For example, you might be allowed 100 dBA over 1 minute, or 100 dBA over 15 minutes. In general, LEQ readings are a better indicator of the total noise exposure than peak readings are.

Ambient Sound Pressure

When you're soundchecking and the SPL meter is displaying its reading, you should be aware that it isn't the same as you will get when an audience is in the room; ambient reflections in an empty room will result in higher readings than you'll get when the room is full. When the room fills up, you'll lose at least a couple of decibels.

Pressure Zone Readings

If you're mixing in a pressure zone and you have the mic from the SPL meter right next to you, you aren't getting a reading that represents what the audience is hearing. As we discussed earlier, the sound pressure in pressure zones is nearly double.

Find Out the Limits

Make sure you find out if there are any limits in place and what they are before you start mixing your soundcheck. You don't want to have to turn down halfway through a song because decreasing the volume isn't a simple case of just turning the whole thing down. Once the volume decreases, the relationships among frequencies and the room begin to change. It has to be a slow process, and you must work backwards in level. Keep in mind, though, that bodies absorb sound—so if you are just around the limit when soundchecking, you'll probably be under it once the audience enters.

FINISHING THE SOUNDCHECK

If you are having trouble with a particular sound or instrument, don't be afraid to get the band to do it again, or ask them to wait until it is sorted out; they shouldn't mind waiting a couple of minutes to make sure they sound good. However, make sure you explain what's going on, and don't make things up if you don't know what's going on. They are like dogs, they can smell the proverbial a mile off, and that doesn't do you any favors. If there is a song that has a certain effect that you need to test, just get the band to play that part of the song. However, you want to try and save the singer's voice, and also you don't want to keep the band there longer than necessary.

One of the best tips I can give you for running a soundcheck is to finish your soundcheck on the first song of the set. This way, you can get all your settings just right for the beginning part of the song. When the band opens their set with this song, the balance is pretty much there, and you need only make a few adjustments, rather than trying to get the mix together. You can then concentrate on refining the EQ and the mix so that it's right for a room full of people.

The bottom line is that it's all about making things easier for yourself. That way, when the pressure is on, you have less to worry about, and you can take the whole thing in stride.

FESTIVALS

Soundchecks (or the lack of) for festivals are a special case, so we want to briefly discuss them here. Technically, you won't get a soundcheck unless you are headlining and insist on it, or if you are the first to go on (and that's just to check the PA). You'll have enough time to set up all the equipment—some of which, like the drums, will be on rolling risers so that they can be wheeled out very quickly. Because they'll be set up on risers, you'll be able to put all your mics in the correct positions. The reason for this setup is that, a lot of the time, you'll only have 20–30 minutes to get your stuff on stage, line check, and then get the band on. (Some mainland European festivals, which are wonderfully run, allow you an hour for setup.)

In some cases, you may be able to use your own console; smaller digital consoles can be easily incorporated into the stage setup for monitors, and occasionally you'll be able to get a FOH console in there as well. If you do take in your own consoles, make sure you take everything in with you. If you have your own monitor console, you'll have to be able to set up all your lines, and then plug into the house multicore. If you're bringing your own FOH console, you should bring your own multicore, but you might have to be on site very early to make sure it's run out. PA companies at festival sites really don't like it when you turn up with only half the gear you need, and expect to borrow the rest from them.

When you're doing all of this, make sure you do a full and proper advance, so that everyone is on the same page. You might even be able to use the house multi. If the festival is using digital consoles and you have a show file for it, send it over in advance so that they can integrate it into their festival show file if needed.

NO TIME FOR SOUNDCHECK?

Sometimes, you aren't going to have time for a soundcheck, especially if you're an opening band. And again, at festivals you almost never get a soundcheck— just a quick line check as you are doing the change over (and in most places you aren't even able to put any noise through the PA; you have to do your monitoring on headphones).

If you *are* able to hear things through the PA, make sure you do; this will give you an indication of how loud you're going to be. There is nothing worse than having the first chord being played sound quieter than the audience. Make sure there is sound through the monitors, and, if there's time for a song, get the band to play a song; if not, good luck. Quick and efficient is the way it should be. If you don't have a line, move on to the next one so that someone can sort it out.

Don't worry if you haven't got the perfect sound right away. Just make sure you have the vocal, which is the most important part of the mix.

SOUNDCHECK DONE!

Once the soundcheck is finished, make sure your settings are saved, or, if you're using an analog console, take note of all your settings. (Using a desk sheet is the best way of doing this, but if you're really clever, get yourself a digital camera and take a picture of all your settings.) Once that's done, go and *spike* the gear; this is where you mark out all stage positions of all the equipment, mic stands, and monitors. This is so that when you come to do the change over, you know exactly where all the equipment needs to go.

You'll also find that the band loves to run off stage and go straight to dinner, and if you are taking too long writing all your settings down, then you are going to be eating alone. . . .

CHAPTER 15
The Mix

INSTINCTIVE BEHAVIOR

Nature

There is something very natural about music—it brings people together. If you love music, you'll likely have met people because of the music you love. This is something that applies to nature as well. Look at songbirds, for example. Their songs help them attract mates; songbirds kept in captivity have no interaction with other songbirds and will never learn their birdsong. Music isn't genetically passed down—it must be learned, and when the song is passed down, it gets changed ever so slightly, so over time the song evolves.

You must understand the natural elements of music and how we respond to it. Once you understand that, you can manipulate the way your audience reacts to it, just by making them listen to a different mix.

Percussive Creatures

It's natural to react physically when something happens to you. Humans respond to bangs and claps—most loud, sharp sounds will get an immediate physical response. When we're very happy about something, we tend to high five each other, punch the air, or jump up and down. When we are backed into a corner, we push out to try and get away.

All these actions are a natural response of our sympathetic nervous system, which is connected to our hearts. At the very core of all of us is our heartbeat. When we're excited, it speeds up; when we're relaxed, it slows down. Heartbeats form a part of who we are and also accentuate the extremes of our emotions; only when we are in a normal, calm state does our heart rate remain stable. Music touches our emotions, so it makes sense that beats and rhythms in music can really affect an audience's reaction to it. As mentioned earlier, percussion was the first form of intentional man-made music, perhaps because of the effect that a loud bang has on our heart.

The Circle of Emotion

Artists on stage need to get a vibe from the audience to enjoy a show, and the audience needs to get a vibe from the artists to enjoy a show. This circle of emotion can be vicious—it can destroy a show, or it can make it the best show you've ever been to. When the band puts their all into a performance, the audience goes nuts—and that gives the band more energy to give back to the audience. It's very common to hear a band comment on the audience after a show, because it truly has a direct impact on their performance. When the audience looks like they aren't enjoying themselves, the band hates it, but when the audience goes wild, singing along, raising the roof after each song, the band loves it (even though they didn't play that well). The audience can get really wrapped up in a show; if the vibe and the energy are there, they aren't worried that there were a few mistakes made because they probably didn't even notice. A live show is all about living in the moment and making those moments connect with the audience so they don't forget it.

MUSICAL DYNAMIC

In this case, we use the word "dynamics" to refer to the quiet and loud parts of songs, and the *dynamic range* refers to the range between these two extremes. Having dynamics in your mix makes everything sound more exciting. The problem, though, is controlling the extremes without destroying the musical dynamics: you want the quiet parts to be quiet but not too quiet, and you want the loud parts to be loud, but not so loud that they deafen everyone in the room.

In general, the band should be able to control their dynamics pretty well, but it is still something you can help with by giving added boosts to choruses, and pulling the band back when the acoustic guitar and vocal are the center of attention.

With electronic bands, on the other hand, this dynamic range is usually missing because the dynamics are already programmed or prerecorded. This can sometimes make the music feel a little sterile, and even a little flat. The musicians on stage don't always have control over the dynamics that are coming out of a backing track, for example—although musicians might be playing conventional analog gear that can have a massive range.

Controlling Musical Dynamics

One good way to control dynamics is to route everything to one VCA, with the exception of the vocal, which sits on its own VCA. This way, when the band is playing a quiet song, you can pull back the whole mix with one fader and let the vocal ring out. Then, when it comes to the climax, you can level the band and the vocals back together again for some added impact. By doing this, you're creating volume changes between the quiet parts and the loud parts—not leveling them out, but creating a bigger gap between them. As a result, the jump into a chorus is more pronounced and contains more energy. Make sure you are doing this in time with the music, though, as turning things up or down out of time will have the opposite effect. Too much of a gap will leave the vocal standing way out on its own, and then when the music is thrust back in again, the music could overpower the vocal.

Ideally, when you're working with a fully live act, you'll find that individual players on stage create their own dynamics, pulling back together and pushing together. Unfortunately, stagecraft takes a long time to build up, so younger bands might need a little more help.

Buildup, Tension, Transfer

There is more to increasing audience energy than just turning the volume up and down in the right places. Specifically, let's look at the *buildup*, which is when you have a song that starts out quietly and then has an extremely high-energy, balls-out ending where the band goes nuts. In these situations, I like to just let the mix sit for a while, having an understated first chorus, and slowly incorporating more and more elements of the power behind the mix as the chorus keeps coming round. Making each chorus as loud as the next can ruin the surprise at the end, and let's face it, we all like surprises.

Next, let's look at the *tension*, which is important for getting the right reaction out of the audience. Within choruses, you'll be able to hear the buildup, and within the section before the climax, you want to hold the mix so that it sounds like it wants to come out, but is being held back. This is *tension*.

After the buildup and the tension comes the *transfer*; this is the release of energy into the room. Getting this right can be quite difficult, especially if you have a band that isn't good at it; it can involve multiple fader movements. The trick, though, is to think ahead. You can move the faders if they aren't being used, and if a musician pauses, you can move their fader without people noticing. It's all about looking out for the extremes—getting air moving from the bottom-end and having the highs play on the tops of your ears.

Consistency and Continuity

Songs on an album often all sound quite different from each other; some may have massive layers of guitars, while others will have a huge amount of effects on the vocals. As a result, it isn't always easy or practical to re-create the sound of an album live, and in many cases trying to re-create the sound of more than one album. If you have a lot of different sounds in each song, especially in the low end, you'll have a hard time maintaining an energy level.

Let's take a sampled kick drum from each song on the album and use it for each song in the set; you are going to be chasing the EQ to maintain the same level of energy. This is happening because chances are the kick drum sounds on the album have been chosen for their sound, not their high-energy impact level at a show. This will make it extremely hard for you to produce a consistent amount of punch throughout the whole show, and you'll notice this in the audience reaction. You'll hardly ever find that kind of extreme example, but it translates through to all the other sounds in the show. Inconsistencies are worse in the low end because of the way they can affect the energy level, whereas inconsistencies in the high end are much easier to deal with and can come across as tuneful.

The key to making a live show work throughout the whole set is making the set consistent and to make sure that there is some form of continuity throughout the show. You should try and wrap the whole show up and present it to the audience. If the sounds are changing, try and do it over the course of the whole show, not between each song; otherwise it could sound pretty bad.

In most live shows you'll be using a lot of traditional equipment, so its settings are easy to maintain. Going back to the kick drum analogy again, when you have found the right kick drum sound, it is pretty much set for the entire show, and you aren't going to have to worry about getting the EQ right for each song.

The problem with programmed or prerecorded material is that it's easy to change the sounds song by song to match the album. This is something that is much harder to control in a live show, as you'll probably be chasing the EQ around a lot. Each song sounds different, so you will be trying to make each song fit in with how the room is sounding that night. With La Roux, I had this problem a lot when I first started. I had to go back into the multitracks we were using and insert a sub kick drum on all the kick drum tracks—this gave the whole set that bit of sub consistency. (We also high-pass filtered the original drum sound and low-passed the new sub kick drum so we wouldn't get any EQ or phasing issues.)

BUILDING BLOCKS OF A MIX
What Makes a Good Mix?

What's the difference between those engineers who are average, good, and amazing? It's all about how they listen and understand the music. There's a lot of things to consider when putting a mix together. Let's look at the fundamental points of putting a mix together.

EQ

There's a lot of EQ that can be done, and only a very small part of the mix is actually done on the faders. It's more than just getting the right balance in level between the instruments; it's also about making them fit around each other. If you ask engineers about what they are doing, they might have to think a little while about it before they can verbalize it. This is because mixing has become a natural flow of creativity, environment, and thought. Your mix is going to go through many stages and be full of tiny instinctive thought processes, creating excitement and electricity through buildup, energy transfer, and sound level.

SOLID AND CLEAR

Having control of your entire mix plays a major part in being a really good engineer. When your mix is completely under control, you won't hear any frequencies suddenly jumping out at you; everything will sit together nicely, and you won't be struggling to hear anything at all. You'll be able to hear everything that is going on, from the little percussion sounds to the well-rounded, heavy-set vocals of your leather-clad front man. And when the bass player hits a note and the whole room fills up with that frequency, it's not a good sound; the seamless flow between the notes is what you should be aiming for. A mistake that a lot of first timers make is that they don't really know what they are listening to. If the singer has a bad mic technique, rather than trying to sort it out, it's left to be dealt with by themselves. They just put up with what they're given. It's not their fault; they just don't know any better.

COMPREHENSION

The key to all of this is to be aware of what kind of music you're mixing and the type of audience who will be listening to it. I worked for quite some time with an artist called Seasick Steve, who had a very simple setup: The kit was a standard four-piece kit, one guitar line, and a vocal. The kick drum, though, wasn't just any kick drum; it was an old marching drum, with real skin over it. By itself, the drum sounded more like hitting a cow in the stomach with a wet fish fillet than a kick drum, and because it was actual skin it didn't have a hole in the front of it. I just mic'ed it using one mic, very minimal EQ, and no noise gate so the whole thing rang out naturally. It sounded awful on its own, but when the other elements of the mix started to be mixed in, the sounds started to make sense, sonically.

> **SEASICK STEVE'S KICK DRUM**
> Because there was no hole in the front skin, I placed a Shure Beta52 about 4–5 inches away from the front head, then used very minimal EQ on the channel, losing 300 Hz about 3–4 dB, with a Q around 1.0, and adding a slight bit of harshness around 2.5–3.15 k. I let it ring naturally, with no gate. Technically, it was a poor-sounding kick drum—but in the context of the mix, it worked better than anything else would have. This is what understanding what you're mixing is all about.

LEAVE YOURSELF SOMEWHERE TO GO

One of the things we looked at in rehearsals was putting the setlist in order on a computer playlist so you could hear how the songs flow between each other, but most importantly where the peaks and drops in energy are going to be. It's so important to get the right energy in the right place at the right time; remember, when you are starting to put a mix together, never work to 100% straight away, and always leave yourself somewhere to go.

NEVER MIX WITH YOUR EYES

Some of us have perfectly good ears on the sides of our heads. The ability to use them is so important to how we do our job and get the best out of the equipment we have.

I see far too many engineers leaning into a computer screen attached to the console, staring at the plugin settings, or making sure that all the faders and EQs are at 0 dB, rather than sorting out the mix. Computer screens can be a distraction from the job at hand. Before we had a screen on consoles, we never knew what the actual value of the fader was, or what the EQ curve on the bass guitar looked like. The fact is, none of this really and truly matters. As long as it sounds right, it is right, no matter how you got there. There might be an easier way and a better way of doing things, but you'll figure it out, and it is that constant learning that makes you good at your job.

TAKE THE ROUGH WITH THE SMOOTH

There are certainly going to be times, however, when you're going to have to make the best out of a bad situation. You aren't always going to get the best mics, PA or mix position, but what we are all paid to do is to make those situations sound the best we can. And it's making those situations sound the best that is most important to the artist we work with. But as with anything, with the right preparation, this should seldom occur.

Understanding What the Artist Wants

You have probably heard stories about how difficult it is to work with some artist and how he or she asked for the moon to be covered in chocolate, attached to a stick, and used as a lollipop for the rest of eternity. The thing is, these artists usually know what it is they want, but have trouble articulating to the people around them exactly what that is; that's why a lot of the time you hear these ridiculous stories.

When presented with statements like:

- "Can you make the bass come out?"
- "I want it to sound more open."
- "It's not crisp enough."
- "It feels like it's in a box."

You can't just have a blank expression on your face.

The problem is we all have different onomatopoeic words for sounds according to how we hear them, so your job here is to try and interpret what these things mean and translate it into a sound. The trick is to really listen, and then when you think you know what they mean, change your desk settings very slowly. Make sure you have a good line of communication going so that you know what you are changing is right and is going in the correct direction. If you can, get the artists to bring some recorded music so they can give you a reference to what they mean.

You'll get some odd requests from time to time, but you should try and go with them. You never know, you might learn something new.

Frequencies and EQ

HOW TO DESCRIBE WHAT YOU ARE HEARING

As I just said, we all have different words to describe a sound, and unfortunately we don't have any industry standard words to mean anything in particular and so are kind of left up to interpretation. I've tried to compile a list of words that are most commonly used, and what they generally mean to us as engineers, but I'm sure they are open to interpretation.

If we describe a sound as being too *bassy, boomy, flappy,* or *subby,* we are saying that the sound has a lot of low-end frequencies in it, from around 180 Hz and below. Subby would refer to the lower end of those frequencies around 80 Hz and below. You might also hear about bass *extension*; this is referring to the decay time of the bass notes. If something has a long extension, it means the bass takes a while to die down. When a sound has punch, we usually mean that there is a peak in the frequency range about 80 Hz–120 Hz, giving us the feeling of being punched; when it is tight, the sound will have the punch, but the lower frequencies might be more controlled. Without these frequencies the opposite sound could be described as having no *guts.*

If a sound has *body, depth,* or is *fat, full, thick,* or *warm,* then we are generally talking about an emphasis in the lower mids, around 160 Hz–400 Hz. And if these frequencies were pulled out, then the sound might be described as *thin* or *hollow.*

Then there is *honky, middy, nasally,* and *woody.* These sounds would be referring to frequencies from 500 Hz–1 KHz, exactly like the type of sound you get when you cup your hands around your mouth and speak. The word *tinny* would be in a frequency range slightly about that, mostly your whole high midrange, just like the sound through a telephone.

If something was described as *aggressive, bitey, bright, crack, edgy, harsh, hard, raw,* or *sharp,* you'd be looking in the area where we get a lot the definition in our hearing, around 2 k–5 k. If those frequencies were working really nice, the sound could be described as *smooth* or *textured,* but if they were to be removed completely the exact opposite might sound quite *dull.*

When a sound is described as *blurred, blurry, cloudy, muddy, muffled, grainy,* or *woolly,* this would tend to mean that the sound has a bad transient response, so it is hard to define what that sound is. The opposites would be words like *clear, detailed, defined,* or *transparent.*

If something were described as having *ambience, breadth, depth, being open* or *spacious,* or *having width,* this would give us the impression of a sound that has a lot of acoustic space. The opposite of this might be the feeling of a sound being kept in a *box* or being *closed.*

ABUNDANCE OF FREQUENCIES

Let's now look at why your mix can easily get cloudy or woody without much effort at all. You have started putting your mix together piece by piece; you add all the sounds in, and then your whole mix has become muddy. What's happened is you have now got an abundance of frequencies you don't want, and you don't know where they came from. This can happen so easily when having too many sounds around the same frequency, so be aware of it—what sounds great individually might not sound so nice when they're all put together. For example, to get a nice thud to a snare, you might boost 200 Hz. Then, to get some nice low-end rumble from your guitars, you might boost 160 Hz. Your bass guitar sits in these frequencies as well, however, and the next thing you know, you're having a party at 180 Hz, where all the frequencies have started to come together. You then go over to your graphic and pull that frequency of the PA, but you don't have those faders on your graphic, so you pull 160 Hz and 200 Hz out to compensate, changing the phase even more. The same goes for removing frequencies; if you keep pulling out the same frequency in each instrument, you'll end up with a hole in your frequency range. Remember, you must always be mindful of where you're adding frequencies and what you're cutting.

IMPORTANCE OF LOW MIDS

Recently, I have noticed a lot of PA systems lacking in the low midrange. This is a really silly thing to do; it will cause you to lose the depth in your drums and bass guitar, as well as the warmth from your guitars and vocals. Some people are fooled into thinking this is okay because they're overpowered by the impact of the sub in the mix. However, even if that initial impact is very powerful, there's nothing to back it up with. Like a bad beer, the initial taste might be great, but then you realize it's quite weak and it leaves a nasty aftertaste in your mouth.

SOLID EQ

First, let's look at the most obvious major percussive instruments in a live show: the kick drum and the snare drum. You should always EQ these two drums as a whole, leaving space for each of them within the other. Scooping midfrequencies out of your kick drum and boosting those same mids in the snare to compensate for the EQ in the other will make them sit with each other. As I said before, I like my snare drum to sound like a shotgun: a deep thud with an aggressive crack that goes right through the body. But the key to this effect is having a tight sound, using plenty of transients but no long extensions on the low frequencies. Watch the reaction in the crowd (we'll get to that a little later).

The toms of the kit sometimes fall into the background of the mix fairly often because they aren't played as often as the kick and snare. I think placing them higher in the mix leads to a more dynamic sounding kit. Then there are the hi hats, which aren't always featured particularly well as a percussive element— but they can be a great way of adding an extra deeper rhythm to your mix. Normally, a high-pass filter is set quite high, so that you only hear the top end. However, if you move the filter down so you begin to hear more of the rhythm and then add a small notch around 8–10 kHz, you can get some real sparkle.

The bass guitar is the link between all things percussive, rhythmical, and melodical. The bass player controls the tightness of the bass for the whole band. It's hard to affect this when mixing because you have no control over how the musician is playing—just make sure he or she understands how to work with the kick drum. If the bass player starts to move out of time with the kick drum, you'll hear the percussiveness in the low-end fall apart.

The EQ on the bass guitar should fit around what the kick and snare drums are doing. I usually end up boosting mids in the bass that are just above the mid boost in the snare to get some percussiveness and clarity in the note being played, but I also like to put in a touch of 80–100 Hz to give a little bit more thud when the notes are played (PA-system permitting, of course).

With a bass guitar, adding high end makes the guitar sound clicky and poppy, and pulling it out makes it sound dull. Adding some of the midfrequencies, somewhere around 600 Hz–1 kHz, brings out the middy thickness. However, these frequencies are important to the clarity of the bass guitar; try removing some frequencies around 600 Hz–1 kHz in other instruments, so you are opening up that space for the bass to sit nicely.

Once you have your drums and bass sorted out you'll then have a great base to sit the rest of your mix on. A lot of the time the melodic parts of the mix will be fighting over the same frequency range, but a bit of complementary EQ will help them all sit nicely together.

VOCAL RANGES

Most singers have a high and a low range, which can require two different EQs. In the high range, the vocal can become a little harsh (which sounds a bit like a ringing in the top of your ears), and in the low range, it can become a little cloudy. The clarity in a vocal comes from around 800 Hz–5 kHz. (These are also some of the most irritating frequencies we hear because we are more sensitive to them than any others.) However, too much 800 Hz can make things sound a little hollow, while 2.5 kHz can give that biting harshness. Don't completely remove any of these frequencies, though—you will lose definition. As much as they sound horrible individually, they are really important.

FREQUENCY AND EQ TRAINING

Try spending a week locked in a room with a graphic EQ, playing your favorite track through it, and using the graphic to mix the track. This will teach you where certain instruments sit in the frequency spectrum and how they can interact with each other.

Remember how we were talking about how the phase of different frequencies affects others in the spectrum, and by boosting and cutting frequencies, you'll be able to hear what happens when the mix becomes muddy or harsh.

Phase Relationships

As we have looked at numerous times throughout the previous chapters, all the sounds you are working with have a phase relationship. Don't be afraid to use the polarity button (often incorrectly called the "phase" button) on your console; you never know when things are going to interact in a way you least expected it. Whenever there are two mics close (intentionally or not) together that could be picking up the same sound, it's worth hitting the polarity button to see if it makes a difference. Try the polarity button on things like drum overheads, or every other brass mic. Also, if you're having trouble with feedback on a vocal, or if it sounds a little distant, it could quite easily be the interaction of the monitors and FOH. Hitting the polarity switch could solve this problem.

Prerecorded Material

An awful lot of artists use prerecorded material these days, and there is a lot you can do with it. There are two ways of mixing this type of material: either you can be careful how you mix it and try to hide the fact that there's a backing track; or the second approach is you can mix it loud and proud, so that it's obvious. I took this second approach with La Roux. A lot of the music was on backing track, apart from the majority of the keyboard lines, and bass lines, and most of the drums. In the record there's a lot of layering of different synths and drums, and the band wanted to reproduce this layering in the live show, but it was just impossible to play it live. The best way to reproduce the sound was to have the whole thing remixed specifically for the live show, which I did myself in the studio.

The great thing for me about creating an album is that there can be a lot of subtlety introduced; with a live mix, though, this subtlety needs to come across in a different way. This is the great thing about having total control of the backing track mix. When you have a PA system capable of creating all the right frequencies at all the right times, you can have a blisteringly powerful mix, with the highest quality audio. Similarly, the quiet parts can take their place in turn, and the subtlety that was in the album can now be translated into extreme dynamics.

FUSING THE MIX

Now that we've had a look at some of the theory behind what makes a good mix, it's time to look at putting the whole mix together.

Creativity

The whole process of mixing a show is very organic; you are linking your environment, subconscious, and conscious mind. A lot of the movements you make are automatic—you just stop thinking about it and just rely on how you perceive the sound. The big question is: How do you get this ability? To begin, you already know how the songs should sound because you've been listening to the music. Then keep in mind how the music makes you feel personally. Then go from there. Once you understand that, the only obstacle in your way is getting to know your way around the equipment.

Mixing is very hands on: It's so important to be able to feel the music through your body and translate that into what is coming out of the speakers. It can be very destructive if you aren't careful, but when you get it right it feels great. But mixing is as much about creativity as it is about technical knowledge. When you are presented with one set of tools, you might do one thing, whereas when presented with another, you might do something entirely different. But you still need to know how to use the kit to get the sound you are after.

How do we define what is right and what is wrong? Music is all about feeling, so let that drive you, but pushing the boat out and failing isn't a bad thing as long as you learn from it. The fundamentals of building a mix are very similar no matter what you are mixing. Picking out the important parts is essential, for example, vocals and melodies, but making sure they are part of the mix and not sticking out too far is also critical. One set of rules for a certain type of music doesn't always apply to a different type of music. If you know the music you know how it should sound.

Focal Point

The focal point of your mix is the lead vocal; everybody in the room will be listening to the main melody. There is often a danger of your hearing it louder than other people because you know it so well—watch out for this. No matter what type of music you're mixing, the vocal should be well defined and smooth, without harshness or too much low end that will muddy it up.

The level of the vocal depends on what you're mixing. For genres like metal and rock music, place the vocal a little further back into the mix, letting the guitars and drums cloud it slightly from time to time. This gives the added perception of loudness. (Your vocal should always be prominent throughout the whole mix, however.)

Energy Transfer

Once you have got the main melodies down, it's time to build the energy. The setlist should have been put together in such a way as to maximize impact and energy, and you should know where these high points are going to be. You won't be able to understand the proper flow of the set until you have worked through it a few times, but the main high and low points should be noted; therefore, structuring your mix to maximize the parts with high-energy impact will help with the dynamics and the energy transfer of the band throughout the whole show.

As you are building your mix, you need to be thinking ahead. You know which point of the set is going to have the most impact on the audience, so building your mix toward that point can help in creating an extra, underlying excitement. Each song builds up to this moment, slightly louder and slightly more dynamic. One trick here is to slowly work through the high end, adding some of the high frequencies back into the room that are now being absorbed by the audience—but not just stopping at smoothing out the high end, adding a little bit of bite to the mix as well. As the show progresses, the ears of your audience will start to become less sensitive to the high end; adding a touch more will give that feeling of distortion and perceived loudness that can be really exciting. Then, when at that chosen moment, the whole mix peaks, the audience will go nuts. For most artists, this will be right at the end of the set, so you'll have a long time to work toward it. The key is to have a point to work toward, and then structure your mix to maximize the parts with high-energy impact and get those dynamics and the energy transferred from the band throughout the whole show working up to that one point.

Having both punch and dynamics is really important in creating this effect. (You also need to have a good band, though—you can put all the punch and dynamics you want in your mix, but if the guys on stage aren't playing well, they won't do much.) *Punch* refers to those frequencies that you really feel in your stomach. Punch is different from sub, which is lower and which you first feel in your feet. Punch is between 80 Hz and 120 Hz. You need to be careful with these frequencies because they very quickly become overpowering and can swamp the mix; this is a telltale sign of a mixer who hasn't been doing the job too long. The trick is to create all this power without making it stand out. You can do this with every type of music, no matter what it is.

The best place to start is the kick drum. Instead of simply boosting certain frequencies, try pulling out somewhere around 250 Hz–350 Hz, wherever feels best. You'll notice a slight drop in level, so boost the gain. Make sure you're listening for the clarity of the beat; if you can't hear the attack, you might want to think about applying some top end. Personally, I like adding 3.15 kHz; it usually sits nicely without affecting too much else.

We are looking to EQ our kick and snare drums together so that they add a solid base to the drive of the mix. As I said before, I like my snares to be tight, so they sound like a shotgun—versus loose, so they sound like a cookie tin. However, always keep in mind what you're mixing. A nice tight snare drum works with most forms of popular music, but with bluegrass, jazz, and that kind of thing, the sound of the snare is integral to the sound of that kind of music. If you've pulled out a frequency between 250 Hz and 350 Hz from the kick drum, you might have space to add these frequencies back into the snare, but you might find going a little lower, to around 200 Hz, works better for the music. You're looking for the same idea as the kick drum: punch and attack. Once the snare and kick are locking into a groove, you should be able to hear and feel that punch really working.

The interaction between the drums and the bass is key to getting the feel of the music and the energy right. The bass guitar is part of the rhythm and should be treated as such. The mix of bass and drums will be a key feature of the music you're mixing. In most popular types of music, the bass lines are run quite loud, so the EQ in these two instruments needs to complement each other. But the bass needs to have its own place to be heard properly, as you won't always be able to hear the bass lines as the notes become lower. The energy from the bass comes in two parts: the punch and the groove. The player creates the groove that the band reacts to, and you create the punch for the audience to enjoy.

Creating Space

In music, they say it's not what's played that's important, but what isn't played. In other words, having the space for the notes to ring out and come together to form solid segments is important in the overall dynamic of a song. The same is true when mixing. Firing everything down the line will just cause you to feel oppressed by a wall of sound. If you're only working with a three-piece band, you may get away with it, but anything more than that and you could be in deep water.

When mixing a live show, everything can come at you all at once, and sometimes it's hard to tell the forest from the trees—especially when you have a complicated mix including string and brass sections. You are going to need to find space for each instrument and each instrumental section to sit in.

There are two ways to create space in your mix: EQ and panning. EQ will help you find the space in the frequencies that you need, and panning will help you find a physical spot along the horizontal plane. One of the big questions in live mixing is whether using the pan control is a good idea; the concern here is that people standing on the left won't be able to hear much, or any, of the right speaker, and vice versa, which is a serious drawback. The perceived balance of the mix would be completely different at each side of the room, and the balance of the mix is more important than the imaging. Only the people standing near the center of the speaker stacks would hear the overall benefit of the stereo panning as you set it up. But all the audience members will hear some spaciousness caused by the instruments you have panned. I believe that it's important for your ears to be able to hear some kind of stereo image; stereo imaging is exciting, and having a mono mix can feel a little oppressive.

We now need to find space for all of the instruments so that nothing clashes. This can sometimes be a mammoth task, and with the interaction of monitors and a potentially loud stage, it can become an even bigger problem. Making a live mix simpler than the mix on the album is important; you don't want to fill up the space in the room with layers and layers of guitars. If you do have a lot of layers of instruments, start simple and work your layers into the mix rather than starting with all of them. Remember that everything has its own place in the frequency spectrum, and just because some instruments are working in the same frequency range as the vocal doesn't mean they can't occupy the same mix.

Work on a section as a whole, whether it is brass, vocals, strings, or guitars. Whenever multiple instruments occupy the same frequency range you'll struggle to get clarity. EQ each instrument so it sounds great, then add the others into the mix. You'll hear certain frequencies getting swamped, which are only the instruments that produce the same frequencies. All you need to do is adjust the EQ slightly so that the whole section sounds like it's full and smooth. Grouping these sections together and then compressing them as a whole can create a solid block to work on. Once you have found the right EQ you can start looking at panning your instruments along the horizontal plane.

Panning can help separate cluttered sections of instruments, but when panning your instruments you should match their visual point on stage. You don't want to have the guitar on the left breaking into their guitar solo and the audio coming out the right speaker. If you are going to be doing a lot of panning, make sure that you center the instrument that is playing the solo; otherwise it'll look and sound very odd.

Drums are always good to pan. Place your kick and snare in the center to drive the mix, your overheads left and right to create space over the whole kit, and your toms slightly across the center. This way the cymbals and toms will be separate from the main beat being driven down the center.

Other sections like your brass, backing vocal, and string sections are also good to pan across the center. This will give them room to breathe and also create a wider sounding mix. If you know that one of the musicians has a solo coming up, make sure you have your finger on the pan control to pop it back in the center just before it starts.

With the right EQ, you can set the sections back in the mix without getting covered by anything else, still audible and sounding full. You are creating more width with these sections so you might find you won't need to push them as loud in the mix, which creates more space for other important instruments.

Adding or Losing

Mixing is very instinctual, but you can't always follow your instincts. For example, you might think that, if you can't hear something, the best thing to do is to turn it up. This is not always the case though. When you start adding and adding to the mix, everything just becomes louder and louder and more distorted. Instead, you should do the *reverse*. Turn down instruments that are too loud It's hard to grasp at first, but it becomes more natural once you get used it. (This works in all aspects of mixing, not just faders, gain, and EQ.) Begin by listening. If the volume is good, listen to where the EQ is sitting; it might just be that you overcompensated when you were tuning the system. Listening and thinking is the order of the day. It takes time to do this instinctively, but once you are in that frame of mind, it'll be easy.

> When La Roux plays "I'm Not Your Toy," there is a section toward the end of the song where a synth solo is played and carries on until the end of the song. The chorus comes back in again with the solo playing and Elly singing over the top. Her vocal clarity sits around 1 kHz–1.6 kHz, which is also where the synth solo sits, and which covers up the vocal. However, pulling out those frequencies on the synth solves the problem and is much better than simply turning the volume up or down.

CREATIVE DYNAMICS

Compressors are complicated, and not a lot of people really understand how to set them properly and use them to their full advantage. However, they are a vital tool in creating any mix, for many reasons. They aren't just for holding off peaks—they can be used to create an extra level of dynamics, or extra under-rhythm, that works with the music you're mixing.

When we discussed compressors earlier, we focused on the functionality of the controls. Now we'll look at what the settings actually do and how to use them in the best way possible. But before we get to the juicy stuff; if the compressor lights aren't moving at all, then you're not compressing anything! You're

> Really good dynamic processors can sometimes be hard to come by, so over the past few years I've bought a few very nice comps that will help me out of any situation. I got hold of an Avalon 737, which is a lovely preamp, compressor, and EQ combo. It has a valve in it, and it can really warm up any sound. I also have a couple of distressors which, when used properly, are utterly amazing. They give you the ability to crank up any input and get the signal overdriven, and the distressor just handles the input level. The transparency of this is mind blowing; you can compress a huge amount without losing any of the extremities of the frequency band and making the sound feel squeezed. I also own a BSS DPR901, which is a dynamic EQ; the best way to describe it is *subtly responsive*.

merely using the compressor as a rather expensive tool for turning down the input signal.

> When I did my stint with Get Cape. Wear Cape. Fly, I tried the Avalon on Sam's vocal, but it didn't work very well; it didn't let the vocal sit in the mix the way I wanted it because it gave it a bit of a retro sound. You can hear the valves working in the Avalon, and when you're dealing with electronic music, it doesn't always sound right. I had always had a problem with the sound of the acoustic guitar: It sounded cheap; the Avalon really brought it to life. We ran it inline, so the signal ran through the Avalon and then the desk channel, rather than the Avalon being inserted into the channel. I then used the BSS DPR901, which is a dynamic EQ, on Sam's vocal. Sam's singing can sometimes be breathy and sibilant, pushing a little more of the low mids; the 901 really smoothed the whole thing out and made his vocal very consistent.

Back in Chapter 9 we looked at what the controls on the front of our dynamic processors are and what they do. Now, we are going to look at how to use them. On a compressor there are only four main knobs you need to worry about: ratio, threshold, attack, and release.

Ratio

Your ratio is a ratio! In other words, don't be fooled into thinking that, if you keep it set at a specific signal point, it will always be the same; rather, it depends on how much signal you have going into the unit. The ratio knob thins out the whole sound. The bigger the ratio, the less signal gets through; the smaller the ratio, the more signal gets through (but it can become a little more uncontrollable in this latter case).

I always start with a higher compression ratio so I can hear what the compressor is doing. When I find the sound I like, I pull the ratio back until the instrument is compressing in a way that works for the music and the player.

Threshold

It's really important that you have the gain set up on your console; otherwise, the gain structure will start changing, and you'll be chasing your tail trying to get the right amount of gain on the channel versus the right amount of compression. When you begin to set up your compressor, you want it to be over-compressing. This is for the same reason you start with the higher ratio. Once the compressor is set up the way you want it, turn the threshold all the way up so the compressor isn't affecting anything at all, then turn it down so the compressor is working at the right points. You don't need to set your compressor so it's compressing all the time; getting the compressor to add that extra little bit of niceness at just the right moment is what you are after here. If you just leave it compressing all the time, you run the risk of losing any dynamics, and the signal will end up sounding flat.

Attack

Think of the attack time as how much of the transient information you want in your signal. Choosing a really fast attack time makes an instrument sound thin; the more you increase the attack time, the more initial attack you get in the notes. Think about a kick drum; there is a lot of depth in the kick, but the attack from the front skin as the beater hits the skin defines each beat. By setting a compressor to a fast attack time, you immediately take away that attack, and the beats start to blend into one.

Release

This is the time it takes for the compressor to return to unity gain after compressing a note or musical phrase; this is where a lot of the creativity and excitement comes from in a compressor. You wouldn't be mistaken for thinking that the compressor needs to return to unity before the next note or beat is struck, but this isn't necessarily true. With the release knob between your fingers, varying the length of release time, you'll start to hear an extra dynamic that can be utilized. Working with the natural rhythm of a bass guitarist or drummer, you can start to make the instruments sound bigger.

The Process

Your attack and release settings work together and should be used together. They are audibly linked; once you start to change one, the other is affected because you are changing the way the instrument is working with the rest of the music. You should always start at the beginning of the waveform and work your way back. Set the threshold and ratio so that your compressor is in a constant state of compression, adjust the attack time, and then the release time. Once you are happy with how the compressor is holding your signal, adjust your ratio until you are happy with the amount of compression; then adjust your threshold until you are happy with when your compression is applied.

Constructive and Destructive Dynamics

There are two ways to use dynamics: constructively (for correction) or destructively (for production).

CONSTRUCTIVE

Constructive dynamics are using gates and compressors in such a way that you are helping to put the mix, or parts of the mix, in a pocket, controlling the level and cleaning up any ambient sound. For example, for a drummer who can't tune his drum properly, a gate is applied to cut out any ring from the kick or the tom and to stop any bleedthrough from other instruments. Inevitably, you'll come across bass players who aren't consistent when they play; you'll have levels going up and down, all over the place and making your life hell when trying to make your mix make sense. Applying a compressor over the bass channel can smooth out the bass line. This kind of compression is instrument-specific, so once it is set, you can leave it for the whole show.

As well as compressing individual channels, you can apply compression to the whole mix. This isn't something I do or particularly recommend, but if you ever work in the wonderful world of Pop you'll find this useful. Inserting a compressor over the main output of your console is a little like sending your studio mix to be mastered. The compressor can give you a really nice sense of production and can give you a very solid mix. Beware, though, that overcompressing the main outputs can remove a lot of the natural dynamics in your mix, causing it to sound very flat. Even though the mix might sound like a great wall of sound, it won't have much space to breathe.

DESTRUCTIVE

Destructive dynamic processing is used for creative effects rather than for controlling dynamic range. Using gates and comps in this way changes the normal behavior of note envelopes, but this isn't necessarily a bad thing. Destructive dynamic processing is song specific, so you will have to adjust them between songs.

We are talking about things like overcompression. As long as it is applied in the right place, and won't make the signal sound dull and fall into the back ground, it can squash the signal so much that it sounds like it's distorting. Using this on vocals or drums can be quite cool.

Another example of destructive dynamics is using a slow attack time on a noise gate for a kick drum, trying to recreate a heartbeat type sound. A favorite of mine is using a slow attack time and slow release time on a compressor inserted over a kick drum. This gives you a kind of pumping sensation, and can really hype up the beat of the track.

There are three really important rules when applying this idea of compression to your mix. First, remember that transient information is important in deciphering what the sound is, so think about what you are trying to achieve at the end of all of this. Second, adjust in the direction of the waveform envelope: Attack first and then release. Third, there is a very fine line between greatness and insignificance.

Happy comping!

USING EFFECTS EFFECTIVELY

Effects are a great way to improve the sound of your mix; they can add depth, mold instruments together, and warm the harshest of vocals. The art is in getting the right vibe of the song and the artist, and to make it sound like it should be there. In a way, they are very much like using dynamics to change the way an instrument sits in the mix and to add an extra under-rhythm to the beat.

Certain key effects make the artist sound like the artist, and this is what you're trying to re-create. Every type of music has its own specific sound and feel, and effects are a massive part of how light, dark, happy, or sad the music sounds. Let me elaborate. When you hear a certain effect like the typical Joe Meek echo, it automatically throws you back to the 1960s, or putting a gated reverb on drums automatically makes any drum kit sound like it was born in 1984. The effects you add must not only make the artist sound like the artist, but also work with the music he or she is playing.

> La Roux uses a lot of chorus on the vocal within the album, but this doesn't always let the vocal stand out in certain songs when they are played live; sometimes chorus thins it out a little. Instead of a chorus effect, I'll use a pitch shift, which, when tuned close to the original note, gives a chorus-type effect, but allows you to thicken up the vocal.

Before we begin, I'm going to lay down some ground rules for using effects:

Always remember this: Repeats (echoes) should never be right on the beat. By doing that, you are just regimenting the music to a rigid beat. Instead, use your ears and do what feels right for the music. As with the release on a compressor, the echoes do not have to fade away before the next hit comes along. This is especially important when working with music that runs to a click track. If you're working with a band that doesn't have a click track, you'll find that the natural flow is nowhere near as regimented as with the click track.

If you're using an effect as an effect and not just as something to enhance a specific sound, use it sparingly. There is nothing more monotonous than listening to the same delay effect for a whole show. Be varied, and don't show all your cards within the first two songs.

The effects applied to the album will be a lot subtler than the ones you'll end up applying at a show, because you haven't got a studio control room to work in. So the little subtle effects that might make the album sound nice and natural just won't work, as the room will start to cover up anything that isn't blatant.

Don't be afraid to use EQ on your effects. You may need to lose some low mid to enhance clarity, or maybe even add a little top to get that sparkle. Using high-pass and low-pass filters to create the desired bandwidth is always a good idea, and then you can utilize the frequencies in the reverb and the source sound.

There are all kinds of stereo effects, and sometimes they can be a little over the top. Play around with making them in mono, and see if that sound works better for you. Reverbs mainly tend to come out in stereo because they are trying to make a wide effect, but shutting them down can also be effective. The same goes for delays. Most of the time, you'll be using delay on a vocal that is panned center, so it doesn't always need to be in stereo.

Never work against the room—you'll lose. It is much bigger and harder than you could ever be. Let the room do what it does naturally and then add to it. Any room can be an acoustically busy place, and adding effects like reverbs and delays that are supposed to give you warmth and depth might just be a futile task. Sometimes you might be fighting just to get the mix out, and shoving a load of reverb on top of that will just cloud the mix and make it impossible to hear. Pick out the key effects that you use on songs, and identify in which part of the set they're used. And remember, if the room adds enough natural reverb for what you want, then don't add any more; just add it for effect to certain parts.

DIFFUSION

You don't just want to add reverb willy-nilly; otherwise you'll just end up in a washy mess. Controlling the way the reverb sits in the room is actually quite easy, and this is to use the diffusion control. Not all reverb units have this level of control, but if you have a half decent one they should. The diffusion controls the space between the simulated reflections inside the reverb unit, so depending on the type of room we are working in, we can make the reverb we want work with that type of room. If there is a lot of space in the room you are in, then you need to have more defined reflections, and if you have a small room, then you want to have less defined reflections. The larger the room, the more natural reverb there is in there, and by using more defined reflections, then you are letting the natural room reverb sit within your simulated one.

Reverbs

With these ground rules laid down, let's look at some reverb techniques that will help your mix become more natural and solid with very little effort.

Natural Room Verb

You want your mix to sound as natural as possible, so when you encounter a room that is unnaturally dead, you need to warm it up. In these situations you'll probably start to add reverb to things you never thought you'd have to. I recommend setting up a reverb called *room verb*, which is a reverb that smooths out the room and adds a thin layer of reverb to the whole mix. A short pre-delay, and normally a verb time of under a second, will be plenty. Apply this to the whole mix, but make sure you send it through the aux section on each channel; never send the main mix to the reverb, as this will result in a lot of feedback. Using this room reverb technique will help you create depth and avoid a one-dimensional mix.

Thickening Verbs

There's a trick to making a reverb sound huge but without it actually being huge. For example, take a snare drum and apply a normal reverb to it—say around 15–20 ms pre-delay and a reverb time of about 1.2 seconds. In this situation, you wouldn't want to make the verb any longer because it would start to cloud things up; but there's still a way to make it sound bigger. First, take the reverb and pan it all the way to the left. Then get another reverb of the same type, adjust the settings of the pre-delay and reverb time ever so slightly, and pan it all the way to the right. Finally, adjust the EQ on each channel so that they sound a little different—and there you have it, one big reverb. By adjusting the reverb times more dramatically, you can get the reverb to fade in one direction or the other.

Dry Vocals

Don't be afraid to have a completely dry vocal; sometimes this works just as well as putting a stunning reverb on it. Alternatively, you can start with a dry vocal and just add hints of verb here and there, to emphasize points in the music.

Kit Verbs

Drums always lend themselves to reverb; reverb adds more depth and thickness to them. In addition, if you're using gates to tighten up the sound of the skins, adding reverb gives you back that extension in the sound of the drum. If you are adding reverb to the drums, then it should sound like it's part of the drum sound. If you can actually notice the reverb it's too loud, unless you are using it for a specific effect.

Gated Verbs

A classic 1980s reverb effect for drums was the gated reverb; the sound defines an era. A gated reverb is a reverb with a noise gate on it, so rather than letting the reverb naturally die away, the gate comes in and cuts it off. Most reverb units have a gated reverb setting that can easily be adjusted.

If you want to push the envelope, try adding this kind of effect to your guitars and vocals, and try using an actual noise gate rather than the one provided in the unit. This way you can side-chain it to the source sound, so that it only opens the reverb up when the artist is singing or playing, and the natural progression of the reverb isn't interrupted.

Delays

Delays are another good way of adding an extra dynamic to your mix. They can be used as an aid or as an elaborate effect. You can sometimes use a delay in place of a reverb in places. Having a delay under 160 ms means your repeats will be close together, and if your feedback is floating around 0%, you can get a very clear source sound, even in a lively room. This will thicken up your vocal or guitar, but won't cloud the clarity.

When using delays, you need to find the groove of the song. Hear it, and tap it. Just using a normal tap delay will get the delay in time with the song and can make it sound more interesting than just a dry vocal. Be careful when tapping in your delay time that you don't get the classic repeat repeat, which is where the singer stops singing and then on the next beat the repeat from the delay comes in, repeating the line that had just been sung. This can make your mix sound like a cheap record.

If you have an echo that sticks out too much regardless of where you place it, try adding reverb to the delay return. This way it'll be pushed to the back of the mix and become like a ghostly delay. Just watch out for your pre-delay on the reverb; otherwise you'll be sending the echo out of the time with the time you have set on the delay unit.

Another little trick that can work quite well to make something sound a little bigger than it actually is to pan the source sound to one side and then use a delay on the same signal panned to the other side. Use a delay time of less than about 20 milliseconds, and—as it is only the one repeat we need—set your feedback to 0%. Because the delay is so short, you should be able to tell that one side is slightly delayed. This is also known as *automatic double tracking*. You can pan both lines back into the center, and you'll hear a fullness appear. This can also work well if you're having feedback issues with your mics at FOH. Pan the instrument, probably a vocal, to the side farthest away, delay it in the same manner we have just spoken about, and pan that to the closest side.

Another fun technique is to send the delay back into itself; this can give you a very reggae and dub kind of feel, but unlike using the feedback control on your delay unit you are adding audio dirt to your signal path. As the signal goes around, it slowly decays until it becomes a noise. Be careful, though, as it's quite easy to overcook. In this case, rather than being nice and controlled, it ends up making a horrendous noise. If you're going to try and do this, make sure you start with a smaller amount of feedback and increase to taste. You'll notice you can get two extra rhythms with this—one from the delay time, and the other from the feedback. If they are slightly out of sync with each other, then the whole sound can feel like it's flowing more.

CONCLUDING THE MIX
Tightening the Mix

Once you get your mix up and running and sounding pretty good, you may find that there isn't much in the way of definition in certain areas. To fix this problem, you need to tighten up the mix. This process is called *tightening* because it corrects a "looseness" of frequencies (or even just one frequency). This can present itself as diminished definition in the lower frequency range, but we also get a lessening of higher frequencies. This kind of high-end looseness can present itself as harshness or a kind of ping sound. Typically, you wouldn't hear higher frequencies as looseness because their interaction in the room is far less than those frequencies in the lower range.

You can use a few different words to describe this looseness, although they aren't necessarily industry standard. We looked at a few of them above when we started talking about frequencies. So when someone starts talking about the mix being woolly, cloudy, hollow, or harsh, you know what they are talking about and more importantly where you concentrate your efforts.

Above all, remember that percussiveness needs to be tight; confusion in the low-end will not give the right amount of energy transfer to get the audience jumping.

Reading the Crowd

The room is also a living breathing entity when it's full of people. You need to be able to react to the crowd. The crowd also has an interaction with you when you are stuck right in the middle of them.

To judge how an audience is reacting to a show, take a look around. Just by paying attention to the people in the venue, you can see how the audience is reacting in certain areas—and if the response is more positive in specific parts of the room, notice what is different there, and use it to improve the mix for other areas. (When possible, of course—this can't always be done.) Your mix can help or hinder this response from the band. You are directly responsible for making the audience hear what's going on to make them give the band what they need. If you think this sounds a little far-fetched, try turning down the kick drum and see what happens. It's very interesting to watch.

It all comes down to what you are mixing. If you are mixing pop to an older audience, you don't want to have a real hard-driving bass, or a vocal buried so far back in the mix that the hi hat is covering it up. You need to push that vocal out there and pull the bass back (but not so much that you lose the drive). By the same token, a metal band playing to an audience in their late teens needs to be loud! Get those drums up, especially the kick and snare, driving bass. That snare needs to sound like a shotgun every time it is hit. In these types of mix environments, you can get away with sinking the vocal back to the same level as the guitars. Create a wall of noise with driving percussion.

Volume

Be wary of sound volume. Over the course of a tour, you might add a lot more decibels to your mix than you realize. Show after show, day after day, your ears become used to the sound level, and to hear that bite in the mix, you need to turn it up slightly.

Your ears really know when something is too loud only when you begin to hear distortion. This distortion needs to be just a small amount, but because it grates on the ear, the sound becomes harder to listen to (maybe without you realizing). The problem is that, these days, with our efficient PA and electronic systems, the re-creation of high sound-pressure levels without distortion is really easy. So if you ask someone to pass you a bottle of water from the cooler, but you can't hear yourself, it's probably time you turned it down a little.

Volume isn't everything, and you have a responsibility for the ears of everyone else in the building, and of course your own hearing. You would like to have a long career after all, wouldn't you?

Mix Power

To get the feeling of any sort of power, you need to be running at least 95 dB SPL. This is the point where the feet start to tap and the bodies start to shake. When you're trying to create vibe and power for the mix, but you have a limit of 95 dB SPL, you need to go about structuring the mix in a different way. This may be another argument for using a point source system over a line array—with the point source system, you get more volume down at the front, where you want the majority of the vibe.

Ask!

If something doesn't feel right or sound right, don't be afraid to ask someone else's opinion. If you are having a problem with a frequency, and you just can't seem to find it, the in-house engineers are often able to help you. They will know if there is some kind of frequency trap in the room, or if the speaker cabinets resonate at that frequency, or something else to the same effect. They know the room, you know the band, and you're both there to help the audience have a great show.

And one final note: Remember who and what you are mixing.

CHAPTER 16
The Show

SETLIST

The set should always be put together so that the songs flow and there isn't a lull in the set. Quite a common mistake is to put in a fast song, then a slow song, then a fast song, then a slow song. The idea behind this pattern is to avoid lumping a load of slow songs together. In practice all this actually does is make the faster upbeat songs come across with less impact, and the whole set never gets going.

As a FOH engineer you aren't there to put the set together; that is really up to the band. But you should be aware of how the set works when it is played. If you think there is too much of a lull, or the energy of the set gets wasted too early, or doesn't pick up fast enough, then consulting the band about any of this wouldn't be a bad idea.

Usually the first few shows of the tour will have various set changes until one can be decided on. The great point about having a solid setlist that is the same every night is that everyone knows where they're at. I know this might sound a little silly, but when you learn a set, the power works so much better. This also has a bonus for your engineering because you learn how the whole set is mixed. Each song in the set will vary, different effects will be used, and the mix will be different. Night after night, you'll learn where the changes work and the best way to mix the song into one another. The variations between different songs become pretty automatic, leaving you in control of making the sound happen.

CHANGE OVER

The change over is the bit in between the acts where there is lots of running around on stage and gear gets moved off and then back onto stage as the crew is setting up for the next band of the evening. So before you grab your setlist and head off to front of house there are a handful for things you need to do.

First, if you have any wireless equipment, make sure they have a fresh set of batteries. This will ensure that they all stay at their optimum working point. Once you have turned them on, leave them on.

Once you know all your wireless kit is working, the previous band should be departing the stage and their crew should be hurriedly placing their equipment back in their cases. Now all your gear should be put back in place if it has been moved, and the mic positions should be checked.

When you've got the stage setup, grab your setlist, a bottle of water, and anything else you want to take to FOH and go fight your way through the crowd. After persuading the security guard at the FOH compound that you really are the band's engineer and you really do know what you are doing, you will be standing in front of the mixing desk. You want to make sure that all the settings that you wrote down after soundcheck have been restored. For those of you fortunate enough to be using an analog console, you'll have to put all the knobs and faders back to where they were. If you are lucky enough to be the headline act, you'll probably have all your own channels, and everything will still be the same as it was. Or if you were fortunate enough to be able to save the settings, you can recall them. Once you've recalled your settings, make sure that everything is still as it should be because sometimes you can find that someone previously has "safed" a channel, and this could cause all sorts of mischief.

Now that that's done and you've plugged your headphones in, it's time to line check again. Make sure all your lines are coming up in the right channels and check any inserts you have plugged into the channels.

Take your time and methodically work through so that you know you haven't missed anything. If something isn't working, let the guys know on stage and move on to something else; this way you aren't wasting valuable time.

Once you've finished the line check, make sure that any intro music is checked and cued; then put your head in the gig.

THE ANTICIPATION

Now you are in the venue with the audience, which hopefully is really excited about the imminent arrival of their favorite band. A problem you will encounter with the audience at any show is that a lot of them have seen the support bands. One of the ear's natural defenses is to start shutting down the extreme high frequencies, and as more and more exposure to loud noises occurs, eventually the ears become insensitive. If you have just arrived at the console after being out for a lovely meal with your band and crew, your ears will probably be a lot more sensitive than the audiences'. So don't be offended if one of them tells you to turn up in the first song or so. And because of the same situation, you only have about the first three songs to get the sound right before your own ears start shutting down.

You need to be focusing on the first moments of the set. For me, the most exciting part about doing a live show is the entrance of the band. All the hard work you put in that day getting everything set up, working, and getting the sound to meld together and sound like a solid unit has built up to this moment. This is where you know that it has all paid off, and this challenge is one that engineers all over the world face every time they press the unmute button.

Have all the channels that you need for the first song on a mute group, or if you are using a VCA, which I do, have that muted so that when you get the go sign you only have to press one button and the channels go live. If you are using a radio mic that comes in from the side of stage, make sure this is the last thing to be unmuted as you could give the crowd an unwanted conversation. Think about making the first few seconds as stress free as possible; then when that moment comes, you don't end up in a panic trying to press play on the intro, muting the faders, and pushing them up.

Now you just have to wait for the green light . . .

THE FIRST SONG

This is the most important song of the whole set as far as you are concerned. This is the song where you hear if all that effort you have put in at soundcheck has actually paid off. This is the entrance for your band; this is what they are presenting to their fans as the excitement starter.

You are now going to hear the glaringly obvious differences between the empty and the full room. You may begin frantically, starting to add the top end that you pulled out the system when the room was empty. You'll also encounter different gains: The band is now playing for the audience, not just you and a handful of bar staff, so there is going to be energy and enthusiasm that wasn't there before. This is what you are trying to capture, so be careful about readjusting the gains; if they are well over the top then do, but you don't want to lose the energy.

Hopefully you'll be lucky enough to start off with a song that slowly introduces all the instruments so that you can apply little changes and get the balance correct one at a time. But that is quite rare, so you'll probably end up with all the instruments coming in at the same time, so concentrate your efforts on getting the vocal out. If you are having problems getting it out enough, then you are probably running something in the vocal range too hard; pull back things like guitars, keys, brass, and strings that can sit and smother vocals. You can always change the EQ slightly to make them sit elsewhere. Once your vocal is out there, move on to the power. Get it pumping, and once it's pumping start making the clarity appear. Slowly and gently improve on your mix, always keeping in mind where you have just come from because if you make the wrong move you'll want to go back.

SET CHANGES

The ability to read the audience is another thing that takes time to establish. It's not that arms in the air, cheering, and jumping about don't give away the fact that they are enjoying the set, but it's the ability to read how they will respond to what's coming up. A lot of artists will have one set and stick to it; some acts will develop a set over a few shows; and then you get the ones that either can't read the setlist because the print is too small and so announce the wrong song, or can't see that the crowd reaction is at a certain point and that dropping a different song in will enhance their experience. So watch out for changes that might happen. Once you get to know the band you'll start to know where they are coming from, and possibly you'll be able to anticipate when a change might occur. And it keeps you fresh and on your toes.

DON'T JUST STAND THERE!

Once you have your mix under control, and there isn't anything protruding out of the mix that shouldn't be there, get over toward the audience and try and have a listen to what it's like for them. It's so important to do that if you are stuck in a corner or on some kind of platform. You should also move on and off center as well, if you can. Sometimes you are just going to be stuck because the venue is so full; going walkies might take you a long time. If you have the opportunity and the space to do it, however, you really should. But always pre-empt what is coming up: You don't want to be away from the console when that all important guitar solo comes up.

THE HEAT IS ON

About 15 to 20 minutes into your show, you might notice your whole mix starts to change. You might have to add some top end here and there, or maybe pull out some low mid to release that little bit more clarity. You're maybe standing on a platform that is raised above the audience; then when you jump off that platform, you may find all the brightness has gone. At the beginning of the book I brush over the idea of how the speed of sound changes with air pressure, humidity, and temperature. Well, this is where we'll cover it in a little more detail.

We now have our show set up, and you are running. The band is giving a great performance, and the crowd that has gathered is loving it, arms in the air, screaming away, jumping up and down. You're having a great gig. A classic FOH engineer's saying is that you need to wait for the band to settle into the show, for the whole mix to calm down. Think about this: You may have noticed that when the audience walked into the venue, they probably didn't walk in all out of breath and sweaty, but when they leave they are. The energy that was created and consequently released by so many bodies into the room causes the temperature and the humidity to rise. The environment that you are mixing in is changing. Depending on the kind of environment and how that environment captures heat and moisture, the mix will be changed in many different ways. You could be in a festival tent, a tiny little sweaty club with no air conditioning, or a massive arena—such environments are completely different, and therefore the factors that are changing your mix are different. No one rule will fix this, but understanding how the mix could change and reacting to it one step at a time will help you keep it under control.

Let's just have a quick look at the physics behind this; it is really simple, and those of you who have been mixing shows before will easily understand what is happening. So we know that sound is transmitted through the air by the molecules contained in it. It's the movement of the molecules that determines how fast the sound moves through them. As the temperature in the room increases, the air expands, which means that the molecules have more room to vibrate and the sound moves faster through it. Another determining factor is the amount of molecules in the air. This is where humidity comes in. As the humidity of the air is increased, the amount of molecules in the air is increased, and in a hot environment it means that there are more molecules to vibrate quickly, so the speed of the audio is increased once more. Increasing the temperature and increasing the humidity cause the air to have more molecules, and they are moving faster than they would at a colder temperature. The more humid the air, the heavier it is, so it stays lower than the less humid air. Because high frequencies don't contain as much power as lower frequencies, when you have an increase in temperature and humidity, the sound will become duller.

Now let's look at acoustics once more. We know that when sound comes into contact with anything it begins to lose some of its power. High frequencies don't contain a lot of energy, so when they come into contact with any thing that will give that slightest bit of resistance they lose some of the energy they are carrying.

Taking this into account, think about the audience and the energy they are expelling. This translates into a layer of hot humid air forming over the audience. This *humidity layer* will sit anywhere between 2 and 5 feet above the heads of the audience. As this layer has a larger density than the air above it, some of the sound is reflected off the top of it. This is where the high end comes in. Because it contains a lot less energy, it can't move the molecules in the air as well as low-range frequencies, so instead of being absorbed by this dense layer of air it is reflected. Hence when you are mixing either on a platform or when you stick your head out above the crowd, you'll hear more high end.

Think about this the next time you are putting a PA up. When you ground stack a PA, you'll probably be pointing it on a 90-degree angle to this thermal layer. The high end, which is usually at the top of the box, will be sticking out about this layer and will skip quite nicely across the top of the audience's heads, like a flat stone over water. Or think back to the story I told you about Glastonbury and the umbrellas. We need to look at penetrating the layer, and the best way to do so is by the angle the audio hits this layer. Like the descent of a space shuttle through the atmosphere, we need to get the angle just right, we want the audio to go directly into the layer, so there is less space between where the audio hits the layer and enters the ears of our audience.

Of course, you need to have a hyped-up audience in the first place, and they all need to be standing close together to create the heat. It's hard to get the same effect at an acoustic concert for the elderly. The other factors involved are

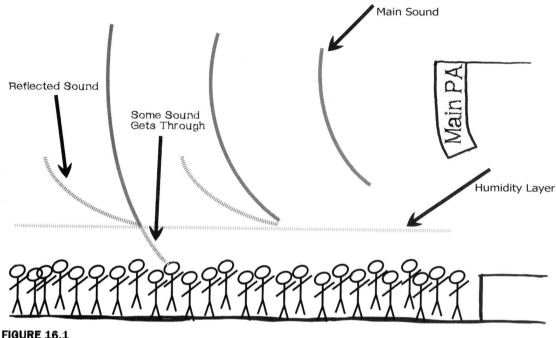

FIGURE 16.1
The Humidity Layer.

how the heat and humidity are trapped. If you are in an open sided tent with a breeze flowing through this will blow away that layer, or if the aircon is turned up full.

So what does this all mean for all the hard work you put in at soundcheck and getting those mic positions right? Well, it's a live show, so there is going to be compromise somewhere along the line. Ultimately we can take control of some elements, but others we have absolutely no control over whatsoever. We can't control our environment; you'll be left with too many questions and with no way to solve them, and believe me I thought about different ways to control the enviromental factors. You can take a temperature and humidity gauge around and record the averages at each show and then try to re-create that during a soundcheck, but then you get left with all the acoustic questions. It's impossible to re-create that environment without the power of the performance and the reaction of the audience.

Every time you mix a show you will encounter your very own unique sound-scape, but one thing you can't take away from this is that when it's cold, the sound is bright and edgy, but when it's hot and muggy the sound becomes softer and more intimate. So now you can understand the age-old engineer saying of "It'll sound better when there are people in," and have a wonderful scientific reason to persuade the management that you are right.

LISTENING

When I'm mixing a show, I rarely listen to the songs I'm mixing; I've heard them before, and I know what is coming up, so nothing is really new. I'm listening so intently to the sound, listening to the individual instruments and their placement, and how their tone complements each other. The other thing that is most important to listen to is how the flow works between each song. This is really important when you are working out how your impact levels are operating. Recall how in Chapter 6 you had put the band's set in order and listened to the energy level of each song. You were listening for the peaks and drops in the set to find where the dynamics of the show were peaking and dropping. Listening to the flow between each of the songs will help you judge your overall level and how the audience is responding to it.

Always listen, but more importantly, understand what you are listening to.

THE END

And there you have it: The last song of the set has been played, you are playing the outro music. There is a massive flurry of local crew on the stage desperately throwing all your gear into the wrong cases in the vain attempt to get all the equipment in the back of the van before the bar next door closes. The merch guy is frantically selling overpriced T-shirts as if there is some kind of worldwide T-shirt shortage, and security is trying to round up the last stragglers standing at the front of the stage who are pleading with the production manager to give them a drum stick. And there you are, fresh from mixing a whole show.

Now it's time to pack it all up and put it in the back of the van, ready to do it all again tomorrow . .

Outro

This industry is still relatively young (even most of my family doesn't really understand what I do—other than follow bands all over the globe on their quest to make money). We are now coming into a far more professional age, where the abuses of old aren't tolerated. If you've made it into the industry, try not to mess it up because you don't know how many chances you'll get. It's a lot of hard work, but look on the bright side: You're getting to see the world at someone else's expense.

Your personality is your most key attribute in any touring environment. Just be yourself; otherwise you'll find it hard to get along—take your own space when you need it. Be aware that you are living out of each other's pockets and that touring on a bus for a month isn't going to be easy. You're all in it together, and there really isn't any room for egos.

It's becoming clearer and clearer to me, as more and more I travel around doing these shows, that more and more in-house crew have been telling us how easy it is to work with us and get great results without having to throw our toys out of the pram. It's not that we try to make things easy; it's just that we know what we have to do and how to do it. Always embrace the problems you encounter, and understand that, chances are, the people you're dealing with haven't exactly caused the problems themselves. Don't make life harder for the engineers who work in the venues—it doesn't do anyone any favors. The next set of guys who walk in through the door will just hear about how much of an ass you've been—they will never hear how great your sound was. Surely you want stories of how great your sound was to become legend, and you want these stories to be told from engineer to engineer.

As we've said again and again, everything in the live audio world is a compromise—the question is *what* to compromise. Even if you have the best PA and mics in the world, there will always be room to adapt. The mic position won't always be exactly where you want it; it's all about the here and now. When you get into trouble, have an order to work in. This is even more important when you're under pressure. Take a step back, and the solution will usually present itself. Your best option in any part of live sound is to learn how to troubleshoot. No matter whether you're the best engineer in the world, or whether you're just starting out, the ability to troubleshoot is always needed (and so is someone who understands how to do a patch).

In this industry, experience counts more than qualifications—but knowledge *and* field experience is the best combination. Be open-minded to new ideas and to new ways of thinking, but don't forget the old. Once you've finished your education, don't think you'll be able to walk out of your college doors and straight onto the tour bus with Radiohead; real work in the field is the only way to work your way through the ranks. Find somewhere to go and push boxes for a few months—then, once you've gained the trust of the people you're working for, you'll be able to get behind the desk. Don't be afraid to ask the people around you for help because they were also in your position once—but remember that they aren't there to mix the show for you. Get your hands dirty and don't be afraid of making mistakes—but make sure you learn from them. Anyone can stand behind a console and push a fader up and down. This is how most people start out, and if you enjoy it, you want to do better and to make it count.

Be ready to work hard. Learn every technical aspect you can. Having a grasp of all aspects of sound will make you a far better engineer. Don't just try and go into a place expecting just to mix; you won't get anywhere that way. You'll have such a narrow focus that it'll take you longer to get to where you want to go. You need to have the technical know-how. If you love it, you'll do anything to do it. You need passion and hunger—because if you don't have both, it's not worth it. The show has to happen, the doors must open—and it doesn't matter if you haven't had a break all day. What does matter are the 2,000 people lining up in the rain outside the venue. They are the ones you are there for, they are the ones who are ultimately paying your wages, so you better do a damn good job and make sure they have a damn good time.

The good thing is that the people you work with closely can become your true friends. This is true even if you're only on a week-long tour—but on longer ones, it's even truer. The people you work with become as close to you as your family, and you'll do anything to help them out.

I came back to the warehouse once, after putting a PA in for a show, it must have been 1:30 a.m. As we pulled around the corner in the 7.5-ton truck Glyn was driving, we noticed something was missing from the front of the building. Glyn's car wasn't there. It had been stolen. We phoned the police and reported the missing vehicle. They informed us that it had been stolen and burnt out!

Whoops.

I had to drive him up to the police station, then waited for him, and drove him home.

You need to have self-confidence and to understand what you're listening to in order to make a show great. When you start out, mix everything and everyone. Try and get behind the desk as much as you can. Mix as many bands as you can and produce as big a variety of music as possible; you'll learn more about music and what sounds typically make up that sound. If you eventually end up in a niche, you'll have the background and understanding to explore your creativity within that niche by possibly bringing in elements from elsewhere.

Remember what you're mixing: Sometimes it needs to sound natural, whereas other times it needs that little something extra. Above all else, never do just what the computer says—make it feel right.

To conclude the book, I'm going to leave you with three golden audio engineering tips. Hold them close, and always remember them when you get behind the mixing console:

- You can't polish a turd, but you can roll it in glitter.
- You're only as good as your last show.
- Above all else . . . trust your ears!

Thanks for reading and happy mixing.

—Dave Swallow
By the pool, West Hollywood, CA,
July 16, 2010

Acknowledgments and Thanks

ACKNOWLEDGMENTS

Tony Andrews—a true inspiration, and Anne
John Newsham, Toby Hunt, and everyone else at Funktion One
Glyn Morgan and all the chaps and Chinnery's and Maple
Mark Saunders and Phil Cummings from Sennheiser
Jason Kelly, Rob Hughes, Richard Ferriday, Simon Moss, and all the guys at Midas and Klark Technic
Ian Laughton, you are definitely one in a million
Joe Wolfe from the University of New South Wales
Peter Lennox from the University of Derby

THANKS

To all La Roux and La Croux; Elly Jackson, Mike Norris, Mickey O'Brien, William Bowerman, Jessica Jackson, Mark Dempsey, Paul Stoney, Risteard Cassidy.

For putting up with my constant need to talk about this book, and bouncing ideas off you, and all your support. You must have been going mad.

Dan Buckley—Miss ya, buddy
Mary Alafetich—Thanks for all your hard work
Tony Beard and all the lovely girls and boys at Big Life Management
Adam Preston, Will Thomas, and Stefan Hensing for all your wonderful work with the images
Stefan Imhof and the hard-working boys at Audio Plus
All the other engineers, tourers I have spent time.
My supportive family; Debby, Ken, Dawn, Georgie, Sandra, Russ, Densil, and Craig
And the two most important people in my life, Miranda and Finn. Thanks for all your love.

FIGURE CREDITS

Adam Preston: Figure 4.1
Courtesy of Audio Plus: Figure 10.6
Courtesy of Midas: Figure 10.7
Courtesy of Funktion One: Figures 9.3, 9.4
Dave Swallow Figures 2.2, 2.3, 2.4, 2.5, 2.6, 2.7, 2.8, 2.9, 4.2, 4.3, 5.1, 6.1, 6.2, 6.3, 6.4, 9.1, 9.5, 9.6, 9.7, 9.8, 10.1, 10.2, 10.3, 10.4, 10.5, 12.2, 13.1, 14.3, 14.4

Iain Furgusson: Figure 9.2
Nick Chmara, VDC, London: Figure 7.1 (and I wrote all over it for you)
One of our drives at the Oya Festival, Oslo, Norway: Figure I.1
Peter Killingbeck: Figures 14.5, 14.6, 14.7, 14.8, 14.9, 14.10, 14.11
Richard Minter: Figure 14.1
Stefan Hensing: Figure 10.5
Front Cover Design by Stefan Hensing and Will Thomas

Index